Real-World Algebra

A unique and understandable approach to functions and modeling

D0209594

Kurt Verderber

Library of Congress Control Number: 2009932799

Printed in the United States of America

ISBN (Student): 978-1-932628-53-1

Software Bundle: 978-1-932628-52-4

Real – World Algebra
A unique and understandable approach to functions and modeling

Dear Student,

This math textbook will be unlike any other math textbook you have encountered! I am making it very different for several reasons, but the most important reason is because <u>I want you to read it</u>. Having taught and talked with thousands of students, and having talked with my professor colleagues who also have taught and talked with thousands more, I can quite positively state that the majority of college algebra students *do not* read their textbooks.

I have never quite understood this, especially since you've probably paid well over $100 for your textbooks in the past, right? Would you ever go into Blockbusters®, rent $100 worth of DVD's, and then *not* watch them? I don't think so…The Number One way to succeed in math is to read the text and do the homework – it's that simple. So why is it that so many students do not read the book? They may be turned off by the math itself, or maybe the writing is a bit dry, or perhaps the pictures are silly, or it could be a combination of all of these things.

That's why I'm writing this book. It's different. I am trying to entertain you and have you read and learn math *at the same time*! That's right…I do not believe that entertainment and math are <u>mutually exclusive events</u> (that means I believe they *can* happen at the same time). They can. And if you read this book, they will!

To be honest, most math texts are written for the professor and not the student. I don't have anything against my fellow math friends and authors, I just never understood why they weren't writing <u>to</u> the student?! I am writing to YOU! And because of that, this text has some properties that you will find interesting. For instance:

First Person…

Throughout the text I will address you as if I were having a face-to-face conversation with you. The use of *I* or *we* let's you know that there is a living, breathing person (or persons) on the other side of the page. There are many in Academia who would say that's inappropriate…I think it's very appropriate. I know that I wrote the text and you know that I wrote the text, so why would I try to hide that fact by using phrases like, 'The student is encouraged to…' or 'The author considers this to be…' I would like this textbook to have a more *personal* feel, and because of that the language tends to be down-to-earth for much of the book.

Stories and narratives…

In some Booklets I have added fictional (or at least partially-fictional) stories to draw you in to the math concepts involved.

Big Word Alerts...

While my main goal is to have you understand college algebra, I also firmly believe in cross-curricular education. In other words, there's no reason why you can't learn some English, science, and history <u>while</u> you are learning math. To that end you will see Big Word Alerts, where I have purposely chosen a large word with the hope that you will take the time to look it up! You will see Acronym Exercises in which you are asked to create different acronyms for those already in place. You should run into some Left brain/Right brain sections in which I try to get you to energize both sides at the same time. There are others...bits of math history, science, etc. All designed to keep you balanced while you study college algebra. You'll be so balanced even Lady Liberty will be jealous!

Unique Applications...

There are some artistic, musical, and numerical applications and scenarios that you might not see in other "traditional" college algebra texts. I hope to show you that these subjects are inextricably linked (hard to untangle), and that math usage is ubiquitous (everywhere!)

In conclusion, I have taken solid teaching practices, wrapped them around college algebra, added a splash of wit and humor, and am now presenting it to you. Read this book and do the exercises – you will learn math and find yourself not liking it a little less.

Sincerely,

Kurt Verderber

The Calculator

To quote Charles Dickens in *A Tale of Two Cities*:

"It was the best of times, it was the worst of times,
it was the age of wisdom, it was the age of foolishness,
it was the epoch of belief, it was the epoch of incredulity,
it was the season of Light, it was the season of Darkness,
it was the spring of hope, it was the winter of despair…"

You may not realize this, but Charles Dickens originally wrote this introductory passage about the calculator. Well, at least I *think* he did. Ok, maybe that's not true at all, but it fits really well…

Let me talk with you briefly about my philosophy of this book regarding how students learn to work with numbers…something we call *numeracy*. Just as IQ is defined as Intelligence Quotient (doesn't quotient really mean divide??), anyway…numeracy could be defined as your Numerical Quotient – how good you are with numbers. One of the main reasons I am writing this text is because my experience has showed me that students enter into college algebra with little or no understanding of how numbers relate to each other.

My experience has been that students do not remember how to add, subtract, multiply, or divide fractions; change fractions to decimals; multiply larger numbers together; carry out long division, and much more. I don't mean to be super-stereotypical here. If you can do those operations, great! You are ahead of the game. If you cannot, hang in there and you will! You must, in fact, because a lack of understanding in numerical operations only gets worse when we move to variables (letters), which can represent any number.

Why? Why do many students have these difficulties with numbers? Well, I'm not Sherlock Holmes (or Charles Dickens), but I think the main culprit is the introduction of the calculator into your 4th or 5th grade math curriculum. The operations that build your mathematical muscle have been stolen from you by that heinous felon – the calculator! At a very young age you were given *Math Drugs*, and you got hooked very quickly. The pressing of buttons replaced the crunching of numbers, your math muscle got soft, and the calculator infiltrated the deepest recesses of your psyche, like Borg nanotechnology, forever corrupting it. As you drift off to sleep at night, calculator safely tucked under your pillow you hear, softly, "resistance is futile…"

You can resist! Ok, perhaps I have over-dramatized the whole thing a bit, eh? And you might be thinking, "Wait a minute! I thought this book uses the TI-83 graphing calculator all the time?!" You're right. We do. The point is *how* do we use it? I am not teaching you to use the calculator for basic mathematical operations…your brain can do that. In fact, your brain is way smarter than your TI. The TI is quite stupid – good at following pre-programmed instructions – but quite stupid. It can't think its way out of a paper bag. What is has on you is pure, raw, speed. The calculator can do hundreds of thousands of calculations every second. We will use it to explore concepts and do those things that would take pages and pages of hand-written calculations to do. The TI-83 is great for Mathematical Modeling, graphing, and things like that.

I will, however, *discourage* you from using it for answering basic arithmetic problems. I intend to build up your math muscle once again with many math push-ups. If used correctly, in balance, the calculator can bring about that "age of wisdom" that Dickens was talking about.

Learning Math as a Foreign Language

Why do a lot of students report a *fear* of math? Can this math-phobia or math anxiety be explained, or is it simply a part of nature?

Is it true that students fear math because they are not necessarily good at math? Or, is the opposite true? Are students not necessarily good at math because they fear math? Maybe this is one of those catch-22's, like, "you can't work here until you have experience, but you can't get experience until you work here."...I don't know...but I tend to think not!

Many of my students have had success approaching math as if trying to learn a foreign language (it is pretty *foreign*, isn't it?) You've probably studied a foreign language in high school or college -- maybe Spanish, French, or Chinese.

Think about these things:

Did you know the language before you started taking the class?...No. That's why you're there. Approach math the same way...it's ok to say, "I don't understand this...I should probably get help, or ask my professor, or get some tutoring." Knowing what you don't know is the first step in assimilating knowledge. Usually, most students start out not knowing what they don't know and it takes them a while to finally know what they don't know. Then they can start knowing what they know...you know?

Did you have to learn the definition of all the vocabulary words?...Yes. In order to use the words to communicate in that language, you simply have to know what they mean. Using the word for 'vacuum' instead of 'hamburger' could be detrimental to your health. The same thing applies with math – you have to learn the math vocabulary. What is a coefficient, a term, or an expression? Just because we are conveying those words in English, doesn't mean we all understand exactly what they mean, where they are used, etc. Try making it a priority to learn the definitions of math words.

Did you have to learn the syntax or construction of the language?...Yes. Sentences in German are sometimes built using inverted word order, with the verb going at the end of a sentence. In German, that's cool! In English, it stupid sounds – see what I mean? Do you know how math words go together? For example, the verb *solve* always goes with the noun *equation*, while the verb *simplify* is used with the noun *expression*. When students tell me they are trying to 'solve the expression', I'm hearing vacuum – not hamburger!

Did the language you studied have different symbols?...Maybe. In Spanish, the question marks are upside down. In German, there are two little dots above some of the vowels (called umlauts…pronounced 'oomlouts'…like oompa – loompa). In French, they have the Cédille (ç), the Circonflexe (ê), and the Accent aigu (é)…silly French kerniggits!! (That's a Monty Python reference, by the way). These symbols mean something; they give you very specific instructions regarding the language. Math has symbols too! In some cases, one symbol can mean an entire English phrase. For example, the math symbol ∈ means "is an element of."

How often did you translate back and forth between English and the foreign language?...Often. Much of learning a foreign language is the practice of translating sentences and phrases back and forth, oftentimes because some languages are *idiomatic*. That means they use idioms (or sayings) which are hard to translate word-for-word. For example, the German phrase "*Morgen stund' hat Gold im Mund,*" when literally translated means "The morning hour has gold in the mouth." Now, unless you are using some seriously expensive toothpaste, I don't know what you're talking about. What it is trying to convey, however, is "The early bird catches the worm." That's how they say it….go figure! Math does the same thing. Math sentences, when translated to English, will look differently, and vice versa. For example, take the English phrase *the sum of three and one-half of a number*. When translated to math, it looks like this…3 + ½(x). Practice, practice, practice!

Language and mathematics are probably more related then you think. The English language is an algebraic language. The construction of English follows certain algebraic properties. How so?

If I told you to <u>buy more soda and chips</u>, what would you buy?...correct, more soda and more chips. So,

more (soda and chips) becomes more soda and more chips.

((watch as we slowly transform…))

m(s and **c)**…becomes **ms** and **mc**

((*and* means +…so))

m(s + c) = **ms + mc**

Viola! What we just saw was an example of the Distributive Property in math played out in English. As we go through the text we'll look at more examples.

Bottom line – Math is a foreign language. If you approach it that way, you may find yourself understanding it better, feeling more comfortable with it, and ultimately (dare I say), *learning* it.

How to use this Textbook (Dear Instructor...)

This text is broken up into *Booklets*, which you could almost equate with chapters. The Booklets typically have one main theme weaving through them, so it seemed more appropriate to name them "little books."

Each Booklet has several *Professor Practice Pause* (PPP) sections which help solidify the previously studied concepts. Typically, these would be completed by students in class. Listed on the bottom of each PPP are corresponding homework assignments from the End of Booklet Exercises. I would recommend that these exercises are assigned as homework when each PPP is completed. There are also *Group Pauses* specifically designed for collaborative work in class. They, too, have corresponding homework problems.

Placed throughout the text are unique applications centering around some of the presented concepts. They are listed in the table of contents in Arial 12 font. For example, in the Booklet on Real Numbers you will find (App) The Golden ratio.

It is my intent that this book be used as a workbook/textbook for the following reasons:

(a) There is enough space provided in all of the *Pauses* and in the End of Booklet Exercises for students to complete the work right on the page.

(b) Almost all of the Topical Summaries are left blank for students to complete. This is a wonderful opportunity for them to review the Booklet, summarize, and write the important concepts in the space provided.

(c) There is also enough white space on many of the pages for students to take notes and write in the margin, as it were.

(d) The text is 3-hole punched for ease of use with a 3-ring binder. This will allow any additional materials from the class (supplements, quizzes, exams, etc) to be kept in one location. Encourage students to do this for organizational purposes.

(e) The pages are perforated for ease of handing-in homework. The 3-ring binder allows students to place pages back in the correct spot after grading.

This book is a consumable item. As such, it may be wise to emphasize to students that it should not be sold back, but kept as a valuable resource for years to come (or at least until the final exam!). Encourage them to write, take notes, underline, etc., directly in the text!

The Booklets are organized based on my experience in the classroom, and the use of concepts in real life. For example, exponential and logarithmic functions are presented before polynomial functions because there are more real-life applications for these two families. However, since the material in these two Booklets is usually more complicated than polynomial functions, you may choose to reorganize the order of the Booklets for your convenience, students, and/or personal style. Similarly, complex numbers are given treatment as a Booklet, instead of being mixed with quadratic functions, as is typical in many texts. Feel free to mix and match! As a consumer of this text, if you ever have any questions, comments, feedback, corrections, and/or suggestions, please email me at *verderk@cobleskill.edu*.

Real – World Algebra
A unique and understandable approach to functions and modeling

TABLE of CONTENTS

Booklet on Real Numbers

The Number Line – Where do you find Infinity?

Would it surprise you to know that if you were to combine all the numbers you know, see, and work with every day – whether it be at college, banking, food shopping, video games, …wherever – they would amount to only slightly more than 0% of all of the *possible* real numbers. So what are real numbers? Can we define them? How do we work with them? Where do they live? I will attempt to help you answer all of these questions.

The following is my definition of a real number…

> A real number is any number that you can think of.

Think of 5 numbers, any 5 numbers and write them down here _____

All 5 of those numbers are real numbers. How do I know? Because you can plot all of their relative positions on what we call the *real line* (or the number line). Plot them below…

$$\longleftarrow \overset{\displaystyle 0}{\rule{8cm}{0.4pt}} \longrightarrow$$

-∞   0   ∞

No matter how big, how small, or whether positive or negative, or how many places come after the decimal point, all of the numbers can find their home between the symbols on the left and right of that number line. What are those symbols? They are the symbols for negative infinity (on the left) and positive infinity (on the right). Notice that a number that is positive has no sign in front of it.

Infinity is *not* a number…infinity is a concept. It is a place that you can approach, but never arrive at. You can't get *they-uh* from *he-yuh*! You cannot add, subtract, multiply, or divide with infinity – it is conceptual. To be perfectly honest, numbers are concepts as well. They were created (or discovered) by man to help him count his rocks and cattle. You can't give me *three*. You can give me three apples, three dollars, and three kisses, but you can't give me just three – just like you can't give me six pieces of *love*…it's <u>intangible</u>.

Warning – big word alert

So infinity is a concept of a concept…wow! Below are some examples of real numbers.

Subsets of real numbers

ROW 1	$0, 1, 5, -10, 1258, -789012764$
ROW 2	$1/2, 2/3, 17/312, .38, -3.141, 5.\overline{3}, 0.278\overline{278}$
ROW 3	$\pi, \sqrt{2}, e, 9.12345678910111213141516\ldots$

The numbers in row 1 are all *integers*. Integers are whole numbers (no pieces) and they can be positive or negative. We give the set of all integers a special letter name, **Z**.

$$\mathbf{Z} = \{\ldots, -3, -2, -1, 0, 1, 2, 3, \ldots\}$$

The numbers in row 2 are all *non-integers* (otherwise known as fractions or decimals). Fractions, believe it or not, have a very specific definition in math. A fraction is a **ratio of two integers**. In other words the number 6.4/9.5 is technically not really a fraction – we don't like to mix the decimals in with the fractions. However, 64/95 is a legitimate fraction.

Notice that the decimal numbers in row 2 either stop at a certain number of decimal places or they repeat indefinitely. The bar over $5.\overline{3}$ means 5.3333333333…Likewise, 0.278278278278…

When row 1 and row 2 are taken together, they form a set of numbers called *rational* numbers. There are actually 2 definitions of rational numbers, a fraction definition and a decimal definition. We have already stated both of them above

Q (the set of all rational numbers) is defined as:
a) all numbers that can be written as the ratio of two integers (fraction)
or
b) a decimal number in which the decimal stops or repeats indefinitely.

That brings us to the strange looking numbers in row 3. All numbers that are not rational are *irrational*. Irrational numbers are all those that cannot be written as the ratio of two integers. The decimal definition of irrational is a number whose decimal places do not stop and do not repeat.

ENGLISH LESSON

The prefix "ir" means not, as in irregular, irretrievable, irrelevant, irreconcilable, and irrational.

Since irrational numbers have a difficult time being represented as fractions and decimals, they usually have symbols or letter representations.

For example, π is the symbol for the number 3.14159265358979323846264338327950288841... The number 2.71828182845904523536028747135266249775724... is represented by e. One that might be more familiar is $\sqrt{2}$ which is typically approximated by 1.414.

Irrational number	Typical approximations
π	3.14
e	2.718
$\sqrt{2}$	1.414
$\sqrt{3}$	1.732

Symbol	Set of all...
N	Naturals ** (Counting)
Z	Integers
Q	Rationals
R	Reals

** The natural (or counting) numbers are all the positive integers. Basically, the numbers we count with such as 1, 2, 3, 4, 5...

Notice there is no symbol for irrational numbers. That's okay, they'll get over it.

You probably remember that π is the ratio of the circumference of a circle to its diameter (C/d) and $\sqrt{2}$ is the length of the hypotenuse of a right, isosceles triangle with sides of length 1, but where does e come from? As we will found out, e is found in many, many, many places in mathematics, sciences, economics, biology, etc, and is referred to as the *natural* base (more on that in the Booklet on Exponential Functions).

Here's an interesting question... "What is the *next* real number after 0?" Think about that for a while. There is an extended question at the end of the Booklet that delves into that question more deeply (extended question #1)

It's time for the...

Professor's Practice Pause

In groups of 2 or 3 take the words and symbols below and organize them into a tree diagram representing the set of real numbers. A tree diagram starts with a heading on top and then branches out into subheadings, and the subheadings branch, and so on...

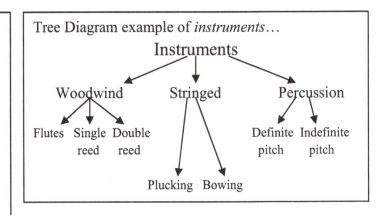

Words / Symbols for your Tree Diagram	Tree Diagram example of *instruments*...
Rational Real Integer Irrational Non-integer Whole Counting (Natural) 0 + and -	Instruments Woodwind — Stringed — Percussion Flutes Single reed Double reed Plucking Bowing Definite pitch Indefinite pitch

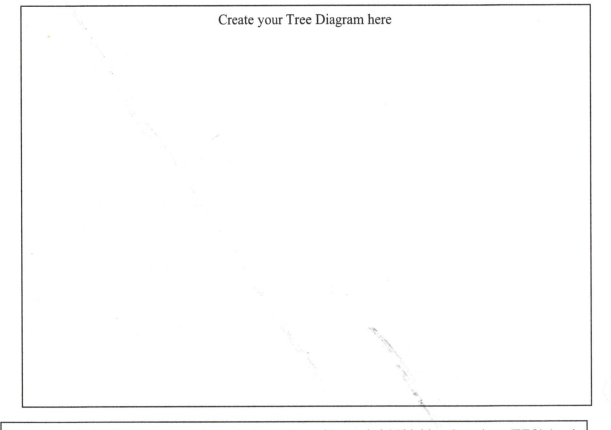

Create your Tree Diagram here

Associated End of Booklet exercises are 2, 3, 6 – 11, and Extended / Thinking Questions (ETQ) 1 – 4

A derivation of the number e

Another math symbol is the exclamation point (!), which is called *factorial*. When the factorial symbol follows a number it tells you to multiply that number by every positive integer preceding it, all the way back to 1.

$$1! = 1$$
$$2! = 2 \cdot 1 = 2$$
$$3! = 3 \cdot 2 \cdot 1 = 6$$
...and so on...
$$8! = 8 \cdot 7 \cdot 6 \cdot 5 \cdot 4 \cdot 3 \cdot 2 \cdot 1 = 40320$$

What if we kept going back to 0? Then all factorials would be 0 BORING!! So...we define **0! = 1**

As you can see, the factorial numbers grow extremely quickly! In fact, the numbers are so large after a while that your TI doesn't even have enough space to store them. Try this...go to the home screen on your TI and enter the number 70. Then press **MATH**, go to **PRB**, and press **4**. You should see "70!" Now press **ENTER**. What you see next is this

```
ERR:OVERFLOW
1█Quit
2:Goto
```

OVERFLOW means not enough space! Like overflowing a cup with coffee.

70! is so large, the number has 100 digits

The number *e* can be represented using factorial numbers. If we set up a sum of the <u>reciprocals</u> of the factorials from 0 onward, we will get closer and closer to *e*.

$$e \sim \frac{1}{0!} + \frac{1}{1!} + \frac{1}{2!} + \frac{1}{3!} + \frac{1}{4!} + \frac{1}{5!} \cdots$$

$$e \sim \frac{1}{1} + \frac{1}{1} + \frac{1}{2\,(1)} + \frac{1}{3\,(2)\,(1)} + \frac{1}{4\,(3)\,(2)\,(1)} + \frac{1}{5\,(4)\,(3)\,(2)\,(1)}$$

$$e \sim 1 + 1 + \frac{1}{2} + \frac{1}{6} + \frac{1}{24} + \frac{1}{120} \sim 2.717$$

Do you remember what *reciprocal* means?

Using only one more term (1/6!) gives an approximation of 2.71806 – remember we said 2.718 is what we typically use for *e*.

There are many, many fascinating numbers out there in the world. The field of number theory is vast! You can study course after course and still only scratch the surface. I find it amazing that there are numbers that are well known and used by many mathematicians, but we still don't know whether they are rational or irrational!

One of the more interesting irrational numbers that has been known and used for thousands of years is the *golden ratio*, represented by the Greek letter φ (phi).

The golden ratio has been studied by some of the greatest minds of modernity and antiquity. I quote Mario Livio at length, from *The Golden Ratio: The Story of Phi, the World's Most Astonishing Number*:

"Some of the greatest mathematical minds of all ages, from Pythagoras and Euclid in ancient Greece, through the medieval Italian mathematician Leonardo of Pisa and the Renaissance astronomer Johannes Kepler, to present-day scientific figures such as Oxford physicist Roger Penrose, have spent endless hours over this simple ratio and its properties. But the fascination with the Golden Ratio is not confined just to mathematicians. Biologists, artists, musicians, historians, architects, psychologists, and even mystics have pondered and debated the basis of its ubiquity and appeal. In fact, it is probably fair to say that the Golden Ratio has inspired thinkers of all disciplines like no other number in the history of mathematics."

Where does it come from? Cut a length into two pieces, a larger and smaller. Call the larger section a and the smaller section b.

If the ratio of the whole length to the larger is the same as the ratio of the larger to the smaller, and if both of those lengths equal φ, then the lengths are said to be sectioned according to the golden ratio.

$$\frac{a+b}{a} = \frac{a}{b} = \varphi \, . \quad \text{where} \quad \varphi = \frac{1+\sqrt{5}}{2} \approx 1.61803\,39887\ldots$$

Leonardo Da Vinci, Salvador Dalí, Mondrian, and many other painters, sculptors, artists, and musicians have used this ratio in the lengths and widths of their paintings, their sculptures, and even in their musical time signatures. Even modern musical groups such as Mudvayne and Tool have been influenced by this number.

Real number Arithmetic (also called Signed Arithmetic)

The four basic arithmetic operations are addition, subtraction, multiplication, and division. Whenever you use these operations with real numbers, you will always get a real number as an answer. In math language, we say that **R** is *closed* under these operations. The real numbers don't let any other numbers join in their reindeer games when they play with the four basic operations.

You have already studied addition and subtraction of real numbers with plus and minus signs (what we call *signed* numbers), but I would like to review one method that I think works well with students. Remember, you should not be using your TI for most of this Booklet!!

USING THE NUMBER LINE (for addition and subtraction)

Every addition and subtraction problem can be done on the number line. How? By thinking of a board game...The first number is where you start (*put your token on start*), the sign in the middle tells you which way to move (+ is to the right, and – is to the left), and the second number tells you how many places to move...it's that simple. Let's look at some examples. Notice that 0 is shown on the number lines as a reference...

Example: Use the number line to find 7 + 3...start at 7, move to the right, 3 units

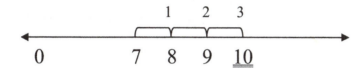

```
              1   2   3
    ┌───────┌──┐┌──┐┌──┐
◄───────────────────────────────►
    0       7   8   9   10
```

Example: Use the number line to find 7 – 3...start at 7, move to the <u>left</u>, 3 units

```
          3   2   1
        ┌──┐┌──┐┌──┐
◄───────────────────────────────►
    0    4   5   6   7
```

Example: Use the number line to find -7 + 3...start at -7, move to the right, 3 units

```
          1   2   3
        ┌──┐┌──┐┌──┐
◄───────────────────────────────►
    -7  -6  -5  -4        0
```

Example: Use the number line to find -7 – 3...start at -7, move to the <u>left</u>, 3 units

```
          3   2   1
        ┌──┐┌──┐┌──┐
◄───────────────────────────────►
    0   -10  -9  -8  -7        0
```

Ok now here's the tricky one. What happens with -7 – (-3)..? In English, "negative seven minus negative three." Remember the card game Uno®? You played the "change direction" card, but then the person after you also played the "change direction" card. What happened? The play came back to you, right?

$$-7 - (-3)$$

Start ↑ ↑ ↑   by 3 units

Go left…no, no, go right

So, -7 – (-3) is the same -7 + 3. *Subtracting a negative is the same as adding a positive.*

English examples of this concept

(a) If you were punished – let's say grounded – stuck in your room, and all of a sudden your mom let you go free, that would be a good thing! Removal (subtraction) of a punishment (negative) is a good thing (positive).

(b) If you owed money to the government and all of a sudden they sent you a letter stating that they had cancelled your debt - that would be an increase in your net worth. Removal (subtraction) of money owed (negative) is an increase (addition) to your net worth (positive).

Also, consider this pattern…

$7 + 3 = \mathbf{10}$
$7 - 3 = \mathbf{4}$
$-7 + 3 = \mathbf{-4}$
$-7 - 3 = \mathbf{-10}$
$-7 - -3 = -7 + 3 = \mathbf{-4}$

Notice that the *value* of the answers (forgetting about the signs) is either 4 or 10. You can't ever combine 7 and 3 and get any other answer, regardless of the sign.

The same technique can be applied to non-integers, but it may easier to use the subtraction technique that you learned in middle school…for example:

$$\begin{array}{r} 5\,1 \\ 2.65 \\ -\ 1.38 \\ \hline 1.27 \end{array}$$

You can't take away 8 from 5, so you borrow a 1 from the 6 (making it 5), and add that 1 to the 5 (making it 15)

Of course, when dealing with irrational numbers, the number line technique becomes more difficult. The answer to $\pi - e$ is anyone's guess, unless we use the approximations, namely $3.14 - 2.718$, then we are back to the technique above.

Notes on a piano

On a piano, if you start at the note C and move upward through one octave of notes (the chromatic scale) you would play C, C#, D, D#, E, F, F#, G, G#, A, A#, and B. Each of these notes is one-half[†] step above the preceding note. (The # symbol after a note means that note is made sharp. C# is read "C sharp". See that, music, much like math, has its own language as well). If you were to start at C again and play downward one octave, you would play C, B, B*b*, A, A*b*, G, G*b*, F, E, E*b*, D, D*b*. (The *b* symbol after a note means that note is made flat. B*b* is read "B flat"). Listing the upper octaves again:

Warning – big word alert

C, C#, D, D#, E, F, F#, G, G#, A, A#, and B

This length of notes is *analogous* to the real number line…it behaves in much the same way. If we assign the number 0 to C, 1 to C#, and so on, we will have:

C	C#	D	D#	E	F	F#	G	G#	A	A#	B
0	1	2	3	4	5	6	7	8	9	10	11

This is an example of a base 12 counting system. We begin at 0 and go up to 11 and then start all over again. There are other bases that are used quite frequently. Base 2 is the language of binary (on or off), and base 64 was also used as a counting system by ancient Babylonians.

To go from C to G#, for example, we are increasing our position from 0 to 8. In math, $0 + 8 = 8$. To go from D# to A#, we increase our position by 7 half-steps, by starting at 3 and going up to 10: $3 + 7 = 10$.

To play down from C to E*b* (using the downward octave) we would start at 0 and go to the left (decreasing) by 9 half-steps: $0 - 9 = -9$. To get from one C to another C two octaves below, we would travel through 24 half-steps: 12 down to the C in the middle and 12 more to the lower C, for a total of 24.

Using this keyboard representation, we see that relative numbers can be treated like distances. In fact in math, we define the **absolute value** of a number to be the distance it is from 0. The symbol for the absolute value of some number, x is $|x|$. Since distance is always a

---

[†]why is it called a half step? This has everything to do with the way the keyboard is set up. When you move from G to A for example (two contiguous white keys) it is called a Step. However, there is a black key in between them (G# or A*b*) and this is the half step. There are two unusual cases…the E key is adjacent to the F, and the B key is adjacent to the C. These are also half-steps because there is no black key between them.

positive variable (except when you get to physics in which case signs indicate direction), we would expect that the absolute value always yields positive values…and it does.

The technical definition is $|x| = \begin{cases} x \text{ for } x \geq 0 \\ -x \text{ for } x < 0 \end{cases}$

> In other words, if the value inside the two lines is positive, you leave it alone. If the value is already negative, you change the sign and make it positive.

Examples: $|7| = 7$ and $|-7| = 7$. The distance from 0 to 7 is 7 units. The distance from 0 to -7 is also 7 units.

$|10 - 12| = |-2| = 2$. The distance from 10 to 12 is 2 units, regardless of where you start.

$|C\# - A| = |A - C\#|$ and they both equal 8 half-steps, regardless of which direction you play. In math, $|1 - 9| = |9 - 1| = 8$.

MULTIPLICATION and DIVISION

You may have been taught to use an x to show multiplication…well, we don't do that in algebra because we use x as our main variable.

Symbols for multiplication include parentheses, asterisk, dot, or nothing at all. The following example shows different ways of representing "five times three":

$$5(3), (5)3, (5)(3), 5{\cdot}3, \text{ or } 5*3$$

You may also have been taught to use this symbol ($\div$) for division, but we typically do not use that one either. Below are examples of "five divided by three":

$$5/3, \text{ or } \frac{5}{3} \quad \text{(notice that fractions are really } \textit{division} \text{ problems)}$$

Again, I expect that the techniques for multiplication and division that you learned in middle school are also applied here…not the TI-83. Build your math muscle by doing that dreaded long division problem!!

What we can and cannot do with real numbers

You are free to add together any real numbers. You are also free to subtract any real numbers and multiply any real numbers. The one operation you are not allowed to do freely (as I'm sure the intractable Mr. Hogthrottle has beaten into you) is division. Well, actually, there is only one division problem which may not be done and that is division by 0.

Mr. Hogthrottle may have said, "Division by 0 is *undefined*!" Now, in my mind, 'undefined' means 'not in the dictionary.' A better definition for division by 0 is *infinite*. Yes, infinite. Ahh, you don't believe me – well I'll show it to you…

Let's approach division by 0 by making the denominator of a fraction closer and closer to 0, as follows:

$$1/5 = 0.2$$
$$1/2 = 0.5$$
$$1/1 = 1$$
$$1/.1 = 10$$
$$1/.01 = 100………What's happening as the bottom gets smaller?$$

$$1/.001 = 1000$$

$$1/.00000000001 = 100,000,000,000$$

As the denominator gets smaller and smaller (i.e. closer and closer to 0), the value of the fraction gets larger and larger (i.e. closer to infinity…∞). Since we can allow the denominator to have as many 0's after the decimal as we want, that means the number of 0's in the answer will continue to grow…So, a better word for division by 0 is <u>infinite</u> – it's more mathematical than undefined, right?

Examples:

(a) 1/0 is infinite

(b) 145,785.23/0 is infinite

(c) 0.0000000000000027/0 is infinite

(d) 12/(3 − 3) = 12/0 which is infinite.

Shoes and shoeboxes…a way to remember 0

Students sometimes have difficulty remembering those two concepts of dividing *into* 0 (like 0/5), and dividing *by* 0 (like 5/0). Think of the top as the shoebox size and the bottom as the shoe size.

So when you see, 0/5, ask yourself, "how many times can a size 5 shoe fit into a size 0 box?"…..none!…0

And when you see 5/0, ask yourself, "how many times can a size 0 shoe (that is, no shoe at all) fit into a size 5 box?"…as many times as possible!…infinite.

If your shovel is empty you can put an infinite amount of shovels-full in the hole and still not fill it!

Professor's Practice Pause

Use the number line method (for addition and subtraction) and the methods you learned in middle school (for multiplication and division) to calculate the answers to these problems...

(a) $-12 - 15$

(b) $6 - 14$

(c) $10 + -23$

(d) $1.5 - 3.5$

(e) $12(37)$

(f) $254(162)$

(g) $125/28$

(h) $2598/14$

(i) $25/0$

(j) $25(0)$

(k) $0 - 5 + 3$

Associated End of Booklet exercises are 1, 5, 21 and ETQ 5 and 6

Fractions, decimals and percentages

There is no such thing as cross-multiplication!! I know, I know, you learned it from the indomitable Mrs. Bumblemeyer, and even used it in math. We want you to forget it! It does not follow any of the proper algebraic transformations of an equation, and students always get confused on when to use it. In my experience, I have found that cross-multiplication almost always leads to confusion. How do I know? Here…take this little, one-question quiz…**Circle** the correct answer.

You would *multiply* two fractions by:

(a) Obtaining a common denominator first.
(b) Multiplying across the top and across the bottom.
(c) Cross-multiplying.
(d) Leave the first fraction, flip the second fraction, and then multiply as in (b)

Before I give you the correct answer, let me tell you an English sentence that will always show you how to multiply two fractions. What's half of fifty cents?...Right…25 cents. So, one-half of one-half of a dollar is a quarter…

"One-half of one-half is one-quarter"

If you can remember this, you can always figure out how to multiply fractions. Let's change the English statement to a math statement.

$$\underline{\text{"One-half}} \text{ of } \underline{\text{one-half}} \underline{\text{ is }} \underline{\text{one-quarter"}}$$

$$\frac{1}{2} \quad ? \quad \frac{1}{2} \quad = \quad \frac{1}{4}$$

The question is, what does the **?** stand for…+, -, ·, or / …?

½ + ½ = 1, so that's not it. And ½ - ½ = 0, so that's not it. Also (½)/(½) = 1, so that's not it either. By the process of elimination, we know that "of" means *multiply*.

So, ½ · ½ = ¼. Now, the only way to get from 1 and 1 to 1 (on the top of the fractions), and from 2 and 2 to 4 (on the bottom of the fractions) is to multiply across the top and multiply across the bottom.

Language note about fractions…

In algebra we *prefer* improper fractions! Writing 11/3 is preferable to writing 3⅔. Why? Remember we said that nothing in between two things implies multiplication? Well, if you saw "3⅔" by itself, without any context, you would not know if it were really 3 *and* ⅔, or 3 *times* ⅔. So, in algebra, improper fractions are always more clear. Thank you and goodnight!

And now…back to our show! The answer to the one-question quiz up above is letter
(b) Multiplying across the top and across the bottom. What about the other answers?

(a) Obtaining a common denominator first:
We do this when we <u>add</u> and <u>subtract</u> fractions with different denominators. The denominator of a fraction is like its unit. The fraction 1/5 means "one of *those* fifths." They're like pizzas. One pizza plus three pizzas is 4 *pizzas*…you need the units.

(d) Leave the first fraction, flip the second fraction, and then multiply as in (b):
We do this when we <u>divide</u> fractions. In fact, we never divide fractions, we change to multiplication. You may recall the phrase "same – change – flip."

(c) Cross-multiplying:
We never do this…ever! *Das ist verboten!* (type this phrase in Google and see what it means)

Operation with Fractions	What to do…
$+$ and $-$	Find a common denominator, then add or subtract the numerators.
$\bullet$	Multiply across the denominator and multiply across the numerator.
$/$	Leave the first (top) fraction, change / to $\cdot$, invert the second (bottom) fraction, and multiply "Same, change, flip"

EXAMPLES

(1) $\dfrac{3}{7} \cdot \dfrac{5}{6} = \dfrac{3\,(5)}{7\,(6)} = \dfrac{15}{42}$ using the principle above…multiply across top and bottom.

(2) $\dfrac{3}{7} / \dfrac{5}{6} = \dfrac{3}{7} \cdot \dfrac{6}{5} = \dfrac{3\,(6)}{7\,(5)} = \dfrac{18}{35}$ using "same, change, flip"

(3) $\dfrac{3}{7} + \dfrac{5}{6} = \dfrac{6}{6} \cdot \dfrac{3}{7} + \dfrac{5}{6} \cdot \dfrac{7}{7} = \dfrac{18}{42} + \dfrac{35}{42} = \dfrac{18+35}{42} = \dfrac{53}{42}$ with a common denominator

Notice that we multiplied each fraction by the opposite denominator to get the *common* denominator. And we multiplied by a form of *one* (6/6 and 7/7) so as not to change either fraction …<u>anything</u> multiplied by 1 is still <u>anything</u>.

Professor's Practice Pause

Without the aid of your calculator and with the aid of your brain, calculate the answers to these problems…

(a) $(2/3)(5/9) =$

(e) $(4/13) + (7/39) =$

(b) $(4/5) / (7/8) =$

(f) $(2/5) - (3/8) =$

(c) $(1/2) / (3/2) =$

(g) $1 - (4/9) + (5/7) =$

(d) $(1/3)(5/14)(3/10) =$

(h) $(9/7) - (2/9) + 2 =$

Associated End of Booklet exercises are 4, 6 and 18

Changing fractions to decimals to percentages and back...

Ok so maybe you don't like fractions. Maybe you prefer decimals instead...that's cool!! You should be able to use either/or, and be fluent in going back and forth between different representations of numbers.

$$\tfrac{1}{2} = 0.5 = 50\% \text{ (fraction, decimal, percentage)}$$

Since fractions are already division problems, it's easy to change them to decimals...just do the division.

$\tfrac{2}{3}$ is really *2 divided by 3*. Write the long division as "3 into 2"

0.66

$3\overline{)2.000}$ 3 does not go into 2, so we write "0" above the two

  0   0 times 3 is 0

 20   2 minus 0 is 2. Bring down the first 0.

           3 goes into 20, 6 times, so we write 6 after the decimal

 18   6 times 3 is 18

 20   20 minus 18 is 2. Bring down the second 0.

           3 goes into 20, 6 times, so we write 6 after the decimal

 18   6 times 3 is 18

 20   20 minus 18 is 2. Bring down the third 0.

Since this is not changing we know that the 6 repeats, and we know that $\tfrac{2}{3} = 0.66666...$or $0.\overline{6}$.

To change a decimal to a percentage we simply multiply by 100. $100(0.6666) = 66.66\%$

$$\tfrac{2}{3} = 0.666... = 66.66\% \text{ (fraction, decimal, percentage)}$$

Notice that for each fractional representation of a number, there is exactly one decimal representation in base 10. It could be represented differently in other bases, however, but for the purpose of this Booklet base 10 is fine.

Changing decimals back to fractions can be difficult, but the easiest way is to divide the number created by removing the decimal by a 1 followed by a number of 0's equal to the number of decimal places in the original decimal. For example,

Change 0.359 into a fraction by writing 359/1000.
Change 0.98 into a fraction by writing 98/100.
Change 0.6 into a fraction by writing 6/10.

It's not always necessary to reduce fractions, but if you need to you can by looking for common divisors. If both top and bottom are even, the common divisor is 2. For example,

98/100 = 49/50...cut both in half
44/80 = 22/40 = 11/20

** If the decimal place doesn't stop, you can't count the number of 0's you need, so you cannot turn the decimal into a fraction and you know then that the number must be irrational!

Changing from...	To...	Do the following...
Fraction	Decimal	Long division
Decimal	Percentage	Multiply by 100 (decimal 2 places right)
Percentage	Decimal	Divide by 100 (decimal 2 places left)
Decimal	Fraction	Divide by 1 and a certain number of 0's ex: 0.12345 = 12345/100000

Some important Fraction ↔ Decimal relationships to know

Here is a short table of comparing certain fractions and their decimal equivalent. These would be good to know…

Fraction...	Decimal...
1/2	.5
1/3, 2/3	$.\overline{3}, .\overline{6}$
1/4, 2/4, 3/4	.25, .5, .75
1/5, 2/5, 3/5, 4/5	.2, .4, .6, .8
1/6, 5/6	$.1\overline{6}, .8\overline{3}$
1/8, 3/8, 5/8, 7/8	.125, .375, .625, .875
1/9, 2/9, 3/9, …,	$.\overline{1}, .\overline{2}, .\overline{3}, …$

This last row of ninths shows something quite interesting…$0.\overline{9}$ is exactly equal to 1. I know, I know, you're thinking 'there's no way that's true!' Let me prove it to you…you can even use your TI if you wish. Set the decimal places to **FLOAT** by doing the following: Go to **MODE**, and arrow down to **FLOAT** and press **ENTER**. Now, you will have many decimal places. To change back, go to **MODE** and arrow over to the number **3** and press **ENTER** to set your TI to 3 decimal places.

On your TI home screen (**2ⁿᵈ QUIT**) you should type 1/9 and then press **ENTER**. You should see:

    1/9   .1111111111

The same pattern will be present with all ninth's. 8/9 is equal to .888888888888… and therefore, 9/9 is equal to .9999999999999… But we also know that 9/9 = 1, so $.\overline{9} = 1$.

Prime Numbers

We used the following sets - rational, irrational, integer, non-integer, whole, and natural (counting) numbers – to break up the real numbers into categories. There are certainly other ways to break up the integers...positive and negative, for example, or odd and even.

There are even more distinctions...

Certain integers are called *perfect numbers* because the sum of their divisors (not including the number itself) adds up to the number. The number 6 is the first perfect number. The *proper* divisors of 6 are 1, 2, and 3, and $1 + 2 + 3 = 6$. The next perfect number is 28. The proper divisors of 28 are 1, 2, 4, 7, and 14, and $1 + 2 + 4 + 7 + 14 = 28$.

Is 36 a perfect number?

Figure that out right in here...

Other integers are called *perfect squares*, because they are the square of another integer. 25 is a perfect square because $25 = 5 \cdot 5$, also written as 5^2 (which we discuss in the Booklet on Basics of the Math Language). Other perfect squares are 4, 16, and 100.

Another way to distinguish the set of (positive) integers is to break them up into prime and composite numbers. A *prime* number is a positive integer whose only divisors are 1 and itself. The number 11 is prime because 11 can only be written as $11 \cdot 1$. The number 24 is *composite* (not prime) because 24 can be written as $24 \cdot 1$, or $12 \cdot 2$, or $8 \cdot 3$, etc.

If a number is prime it is not composite, and vice versa, so the set of all positive integers is equal to the set of all primes plus the set of all composites. Well...almost! The number 1 is neither prime nor composite, it is the *unit* number (also called unity).

What about even numbers? All even numbers are composite because they are all divisible by 2 - - except for 2 itself, which can only be written as $2 \cdot 1$. The number 2 is the first prime number and the only even prime number.

Professor's Practice Pause

Take some time, without the aid of your calculator, to figure out and write all of the prime numbers from 1 to 100. What!!! "What are you nuts?!" you say. No, it's not that bad – here are some tips…

(1) Any number divisible by 2 (except 2) is composite. That knocks out 49 numbers.

(2) Any number (> 5) that ends in 5 or 0 is composite because it is divisible by 5.

(3) A whole number is divisible by 9 if the sum of all its digits is divisible by 9. For example, 8172 is divisible by 9 because $8 + 1 + 7 + 2 = 18$ which is $9 \cdot 2$. That means that 8172 is composite (not prime).

(4) There are 25 answers.

You can use this list below to cross out all the composite numbers. I have already started for you

~~1~~	2	3	~~4~~	5	~~6~~	7	8	9	10	11	12	13	14	15	16	17	18	19	20
21	22	23	24	25	26	27	28	29	30	31	32	33	34	35	36	37			
38	39	40	41	42	43	44	45	46	47	48	49	50	51	52	53	54			
55	56	57	58	59	60	61	62	63	64	65	66	67	68	69	70	71			
72	73	74	75	76	77	78	79	80	81	82	83	84	85	86	87	88			
89	90	91	92	93	94	95	96	97	98	99	100								

Use this space for any calculations…

Associated End of Booklet exercises are 12, 13, 19, and 20

Topical Summary of Real Numbers

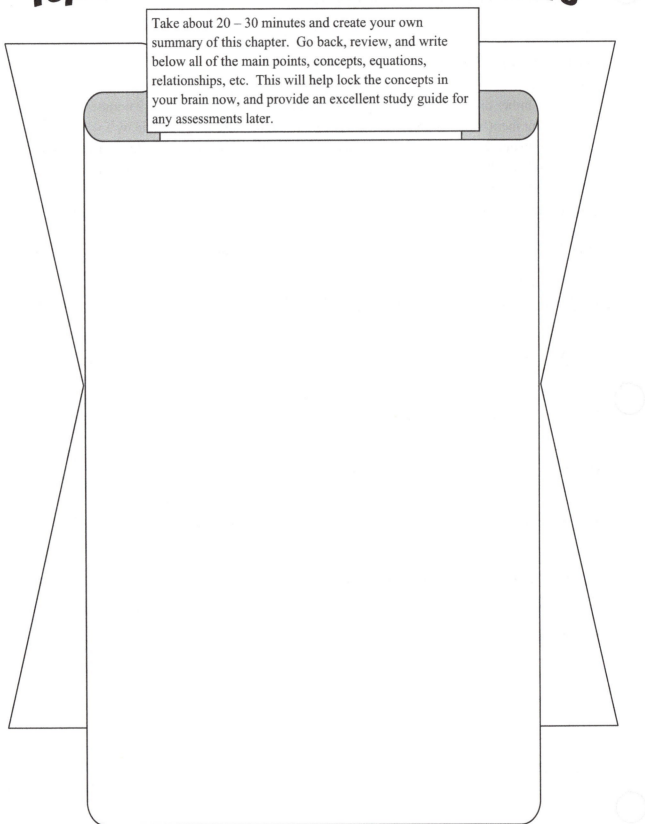

Take about 20 – 30 minutes and create your own summary of this chapter. Go back, review, and write below all of the main points, concepts, equations, relationships, etc. This will help lock the concepts in your brain now, and provide an excellent study guide for any assessments later.

End of Booklet Exercises (the dreaded *EBE*)

1) Without the use of a calculator, fill in the following 12 by 12 multiplication table. A few examples have been completed for you…

	1	2	3	4	5	6	7	8	9	10	11	12
1	1	2										
2		4										
3				12								
4												
5												
6												
7												
8												
9												
10												
11												
12												144

2) Given the following numbers: 8.5, π, 0, -1, -6.75, 32/7, e, -4/3.

   (a) Place the numbers above in their relative positions on a real number line from -10 to 10.

-10   10

(b) Also, consider the following words: real, rational, irrational, integer, counting (or natural). For each of the numbers below, write all of the words that apply to that number. Each number has at least 2 words associated with it.

8.5 _____

π _____

0 _____

-1 _____

-6.75 _____

32/7 _____

e _____

-4/3 _____

3) Can an integer ever be irrational? Explain.

4) Your little brother says that the number 6.25/78 is irrational. Is he right or wrong? Explain.

5)
(a) You continue to divide a certain real number (other than 0) in half. Will there ever come a point at which you cannot take one-half of that number? Explain.

(b) You continue to divide a certain real number (other than 0) in thirds. Will there ever come a point at which you cannot take one-third of that number? Explain.

(c) You continue to divide a certain real number (other than 0) in tenths. Will there ever come a point at which you cannot take one-tenth of that number? Explain.

(d) What can you deduce from the previous 3 questions about dividing real numbers?

5) The square root of every odd integer is irrational, unless the integer happens to be a perfect square. For example, even though 25 is an odd number, √25 is rational since √25 = 5, which is rational (5/1 is rational). Given the square roots of the first 20 integers, that is √1…√20, state which are irrational.

7) List four different elements for each of the following sets:

Q_____

N_____

Z_____

R_____

8) Write three different numbers that are rational but not integers.

9) Write three different numbers that are both integers and positive.

10) Write five different rational numbers between 0 and 1.

11) Write five different irrational numbers between 0 and 10.

12) Why can a perfect square never be a prime number?

13) State two quick reasons why 120 cannot be a prime number?

14) **True or False** State whether each sentence is True or False. Explain your answers using either words, examples, or counter-examples when appropriate.

    (a) All numbers that live on the number line are called *real* numbers.

    (b) 0 is neither rational nor irrational.

    (c) 0 is neither positive nor negative.

    (d) The square root of every positive integer is rational.

    (e) The square root of any negative numbers is not a real number.

    (f) The absolute value of a negative number is a positive number.

    (g) The absolute value of a positive number is a negative number.

    (h) 1 is the first counting (natural) number.

(i) 10 is the last counting (natural) number.

(j) All natural numbers are integers.

(k) All integers are rational.

(l) All rational numbers are real.

(m) All real numbers are irrational.

(n) An irrational number is a decimal that either stops or repeats indefinitely.

(o) 8 is a perfect number because $4 \cdot 2 \cdot 1 = 8$

(p) Two negatives always make a positive.

(q) $3! = 3 + 2 + 1$.

(r) Perfect numbers can never be prime numbers.

15) Perform the following signed arithmetic problems without the use of your calculator:

(a) 3 – 5

(b) 5 – 3

(c) 3 – (-5)

(d) -5 – 3

(e) (-7)(-8)

(f) 16/(-12)

(g) 65 – 16 – 2

(h) 100 – 27 + 48

(i) 10·-3·-5

(j) $\dfrac{75}{-5\,(3)}$

(k) $1.50 – $0.47

(l) 8 – 2 + 12 – 7

(m) 15/0

(n) 0/15

(o) You have five dollars and thirty-eight cents, but you owe your sister three dollars and fifty-six cents. How much money do you have left?

(p) You accidentally drop your wedding ring from the top of the Alpspitze to the bottom of the lake below. The Alpspitze is 2620m above sea level and the lake is 35m below sea level. Donning your scuba gear, you proceed to collect your ring. How many meters did you travel? Given that 1 meter = 3.28 ft, how many feet did you travel?

16) Perform the following additions and subtractions by changing to improper fractions (when necessary) and finding a common denominator:

(a) $2/3 + 5/7 =$ (b) $3/5 - 4/9 =$

(c) $13/4 + 27/9 =$ (d) $3\frac{5}{8} - 2\frac{2}{3} =$

(e) $1\frac{1}{2} + 3\frac{3}{4} =$ (f) $-8\frac{1}{3} - 2\frac{1}{2} =$

(g) $5/6 + 7/8 =$ (h) $9/2 - 23/3 =$

(i) $-9/11 + 7/6 =$ (j) $0/1 - 0/2 =$

17) Evaluate the following expression without the use of a calculator:

(a) $|10|$ (b) $|-10|$

(c) $|5 - 10|$

(d) $|10 - 5|$

(e) $(-3) \cdot |-3|$

(f) $|6 - 5| \cdot |5 - 6|$

Given the following table that assigns the notes on a piano to positive integers:

C	C#	D	D#	E	F	F#	G	G#	A	A#	B
0	1	2	3	4	5	6	7	8	9	10	11

(g) $|D - A\#|$

(h) $|C - B|$

(i) Write the difference between G# and C# in absolute value notation.

18)

MULTIPLY each row by each column   DIVIDE each row by each column

	1/2	2/3	-9/16	24/5	15/18	9/10
0	0·1/2 = **0**			0/(24/5) = **0**		
2/5		2/5·2/3= **4/15**			(2/5)/(15/18)= (2/5)(18/15)= 36/75=**12/25**	
4/3						
3/8						
9/16						
1						
-4/5						
-7/16						

19) Change the following fractions to (a) decimal and then to (b) percentages.

(a) $\dfrac{1}{2}$ (b) $\dfrac{2}{3}$

(c) $\dfrac{5}{8}$ (d) $\dfrac{7}{9}$

(e) $\dfrac{12}{20}$ (f) 1/6

20) Change the following percentages to (a) decimal and then to (b) the most reduced fraction

(a) 5%   (b) 10%

(c) 25%   (d) 33.33%

(e) 77%   (f) 99%

21) Your best friend is an excellent artist, but he's not so good with math. He wants to re-paint his masterpiece (shown below) in the proportions of the *golden ratio*. If the width of his painting is going to be 16", what should the height be? (Use 1.618 to approximate φ)

Extended and *Thinking* Questions

(1) What is the *next* real number greater than 0? Explain your answer.

(2) If you had an infinite amount of time, would it be possible to *count* **Z**? (What I mean by *count* is to assign a counting number (**N**) to all of the integers both positive and negative). How might this be done? Explain or draw a picture.

(3) In a similar manner, would you be able to count **Q**?

(4) In a similar manner, would you be able to count **R**? This question goes hand-in-hand with question (1).

(5) The *Fibonacci* sequence. In mathematics, the Fibonacci numbers are a sequence of numbers named after Leonardo of Pisa (known as Fibonacci), who wrote a book entitled *Liber Abaci* in 1202. It was in this book that he introduced the following sequence:

$$1, 1, 2, 3, 5, 8, 13, 21, 34, 55, 89, 144, 233, \ldots$$

The first two numbers are 1 and 1, and the next number in the sequence is always found by adding together the two previous numbers. The next number above would be 144 + 233, or 377.

   (a) What are the next 5 numbers in the Fibonacci sequence?

   (b) Here's something that might blow your mind!! Look at successive ratios in the Fibonacci sequence. That is, calculate the next number divided by the one before it. For example, $1/1 = 1$; $2/1 = 2$; $3/2 = 1.5$; $5/3 = 1.\overline{6}$; $8/5 = 1.6$; $13/8 = 1.625$. Without the use of a brain-freezing calculator, compute these three ratios. That is, find…

   $21/13 =$

   $34/21 =$

   $233/144 =$

   What number that we have talked about do the divisions seem to be approaching?

(6) We showed that numbers above 70! are too large for the TI to store. Again let me show you that the brain is much smarter than the calculator. Even though these numbers are so big, I bet you could figure out 100!/95! very quickly. Let me give you a hint with a slightly smaller problem.

$$\frac{10!}{8!} = \frac{10\,(9)\,(8)\,(7)\,(6)\,(5)\,(4)\,(3)\,(2)\,(1)}{8\,(7)\,(6)\,(5)\,(4)\,(3)\,(2)\,(1)}$$ and if we cancel out the same numbers in both numerator and denominator we have, $\dfrac{10 \cdot 9 \cdot \cancel{8 \cdot 7 \cdot 6 \cdot 5 \cdot 4 \cdot 3 \cdot 2 \cdot 1}}{\cancel{8 \cdot 7 \cdot 6 \cdot 5 \cdot 4 \cdot 3 \cdot 2 \cdot 1}} = 10 \cdot 9 = 90$

   Now that you understand that pattern, write 100!/95! an easier way.

Research Questions:…to build upon other's knowledge and studies.

(1) Golden Ratio

The *Mona Lisa* or La Gioconda (the laughing one) is probably the most notable painting of all time. Painted by Leonardo da Vinci around 1505, it hangs in the Musée du Louvre in Paris. *The Roses of Heliogabalus* was painted by Sir Lawrence Alma-Tadema in 1888. Research the dimensions of these paintings in centimeters, and check to see if they are proportioned after the golden ratio.

Mona Lisa   The Roses of Heliogabalus

(2) It is not known whether the numbers 2^e, π^e, or $\pi^{\sqrt{2}}$ are irrational. Research some other numbers for which it is unknown whether or not they are irrational.

(3) The *Fibonacci* sequence. In mathematics, the Fibonacci numbers are a sequence of numbers named after Leonardo of Pisa (known as Fibonacci), who wrote a book entitled *Liber Abaci* in 1202. It was in this book that he introduced the following sequence:

$$1, 1, 2, 3, 5, 8, 13, 21, 34, 55, 89, 144, 233, \ldots$$

The first two numbers are 1 and 1, and the next number in the sequence is always found by adding together the two previous numbers.

The Fibonacci can be drawn using squares. Start with two squares of length 1 side by side and then draw the square of length 2 on top…

Since the height is now 3, draw the square of length 3 on the right hand side…

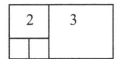

Now, next to the bottom, left, and then top, we will spiral around and draw squares of lengths 5, 8, and 13 respectively…

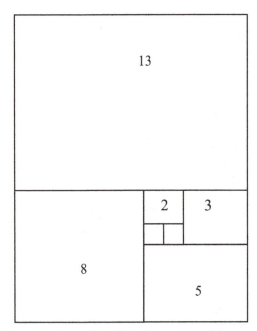

Notice that the sizes of squares so far have been

1, 1, 2, 3, 5, 8, and 13

The Fibonacci sequence.

By connecting the opposite ends of each square with a quarter circle, we create a spiral as shown below…

This spiral is also called Archimedes spiral and shows up in various places in nature, to include sea shells and the arrangement of seeds on some flowering plants. Research some areas of art or nature in which the Fibonacci sequence appears. Some hints are honeybees, pine cones, and sea shells. Write about what you have found.

(4) Look up the *Copeland-Erdös* constant. State what it is. Is this number rational or irrational?

(5) Many mathematicians and numerologists have been looking for patterns in π for quite some time. Our studies would lead us to believe that there are no such patterns in π because of its irrational nature. Research some of the people who have studied π to find "patterns", and explain what they have said or found.

(6) In one of the Professor Practice Pauses, you found all the prime numbers less than 100. On the Internet, look up the *Sieve of Eratosthenes,* explain it in a few sentences, and comment if that method would have been easier.

Booklet on Basics of the Math Language

Exponents...Powers and Roots

You may have noticed that I don't like forcing rules upon students. I would rather teach you how to derive the rules that you need instead of memorizing them. In my experience, memorization leads to confusion (see my discussion about "cross-multiplication" in the Booklet on Real Numbers).

What is an exponent?

An exponent is a little number that sits above the other numbers and letters (also called a *superscript*). It's like the little angel or devil perched on the unwary dude's shoulder, trying to persuade him to do either good or bad. Fortunately enough for our *dude*, there are no choices with exponents – they always behave the same way!

An exponent tells you how many times to multiply the thing under it (the dude) by itself.

$\text{Dude}^2 = (\text{Dude})(\text{Dude})$
$\text{Dude}^3 = (\text{Dude})(\text{Dude})(\text{Dude})$, and so on.

ENGLISH LESSON

The caret symbol (^) can also be used to show exponents.
For example, $10\text{^}4 = 10^4$.

"Ten squared" = "Ten raised to the **power** of 2" = $10^2 = 10\text{^}2$

In our Dude examples above, dude represents what we call the **base**. A nice way of *deriving most of the exponent rules* is to follow a simple pattern...

Start with an easy base (like 10) and raise the base to an exponent that changes from 4 to -4, going down by 1. For example:

$$10^4 = 10,000$$
$$10^3 = 1,000$$
$$10^2 = 100.........$$

Q: how are we getting from one answer to the one below it?

A:....exactly...dividing by 10 (removing one 0)

Let's continue the pattern a bit further...

$$10\text{^}4 = 10,000$$
$$10\text{^}3 = 1,000$$
$$10\text{^}2 = 100$$
$$\mathbf{10\text{^}1 = 10}$$

This tells us that anything raised to the 1 power is itself ... $x^1 = x$

Continue the pattern once more…

$$10^0 = 1$$ ⟶ | This tells us that anything raised to the 0 power is 1 … $x^0 = 1$ |

……..ok…keep going…

$$10^{-1} = 0.1 \text{ or } 1/10^1$$
$$10^{-2} = 0.01 \text{ or } 1/100 \text{ or } 1/10^2$$
$$10^{-3} = 0.001 \text{ or } 1/1000 \text{ or } 1/10^3$$

These last few (with negative exponents) are showing us that "something raised to the negative n is one over something raised to the positive n,"…or $x^{(-n)} = \dfrac{1}{x^n}$

Some numerical examples:

Without the use of a calculator, let's evaluate these exponential numbers:

(a) 5^3 (b) 2^{-4} (c) $(-3)^0$ (d) $(-4)^4$ (e) $(1/2)^{-3}$ (f) 0^6

(a) $5^3 = 5*5*5 = 25*5 = 125$

(b) $2^{-4} = 1/2^4 = 1/(2*2*2*2) = 1/16$

(c) $(-3)^0 = 1$ because anything raised to the 0 power is 1.

(d) $(-4)^4 = (-4)(-4)(-4)(-4) = (-)(-)(-)(-)*(4)(4)(4)(4) = +(16)(16) = 256$

(e) $(1/2)^{-3} = 1/(1/2)^3 = 1/[(1/2)(1/2)(1/2)] = 1/(1/8) = 8$

(f) $0^6 = 0*0*0*0*0*0 = 0$. 0 raised to any power is 0.

How do we combine exponents? I'll lay out the rules first.

Math Operation	Math Answer	English Sentence	Example
$X^m \cdot X^n$	$X^{(m+n)}$	Multiplying exponents with the same base means you add the exponents	$P^2 \cdot P^5 = P^{(2+5)} = P^7$
X^m / X^n	$X^{(m-n)}$	Dividing exponents with the same base means you subtract the exponents	$P^4 / P^3 = P^{(4-3)} = P^1$
$(X^m)^n$	$X^{(m \cdot n)}$	Raising one exponent to another means that you multiply exponents	$(P^4)^5 = P^{(4 \cdot 5)} = P^{20}$
$(XY)^m$	$X^m Y^m$	Exponents can be distributed when bases are multiplied	$(P Q)^5 = P^5 Q^5$
$(X/Y)^m$	X^m / Y^m	Exponents can be distributed when bases are divided	$(P/Q)^5 = P^5 / Q^5$
$X^m + X^n$	$X^m + X^n$	This is not a rule. You cannot combine exponents when bases are added.	$P^3 + P^8 = P^3 + P^8$

If you like to memorize mathematical rules...great! There are a lot of students I know from personal experience that absolutely don't like it or have a hard time with it. Let me show you how you can always derive these rules – all you need to remember is the definition of an exponent. Johnny, the reminder please...

Sure thing Pat...**An exponent tells you how many times to multiply the thing under it by itself.**

Armed with this definition, let's break down $x^3 \cdot x^4$. What we do is write-out the x's...

$x^3 \cdot x^4 = (x \cdot x \cdot x) \cdot (x \cdot x \cdot x \cdot x) =$ how many x's?...7, so the answer is x^7.

What about x^8 / x^3? Write out the x's... $\dfrac{x \cdot x \cdot x \cdot x \cdot x \cdot x \cdot x \cdot x}{x \cdot x \cdot x}$

We can cancel an x on top with an x on the bottom...... $\dfrac{\cancel{x} \cdot \cancel{x} \cdot \cancel{x} \cdot x \cdot x \cdot x \cdot x \cdot x}{\cancel{x} \cdot \cancel{x} \cdot \cancel{x}}$ How many are left?...x^5

How about this rule...$(x^2)^5$? Write out the x's. Remember the 5 means to write out five separate x-squared...

$$(x^2)\,(x^2)\,(x^2)\,(x^2)\,(x^2) = (x \cdot x)\,(x \cdot x)\,(x \cdot x)\,(x \cdot x)\,(x \cdot x) = x \cdot x \cdot x \cdot x \cdot x \cdot x \cdot x \cdot x \cdot x \cdot x = x^{10}$$

Example: Simplify the following expression into one exponent:

$$\frac{x^2\, x^7\, x\, x^3}{x^2\, x^5\, x^4} = x^?$$ To find the ?, add up all the x's on the top and all the x's on the bottom.

By counting we can simplify this expression to $\dfrac{x^{13}}{x^{11}}$ which equals x^2 because there are two left over on top when you cancel the 11 x's from the bottom.

What if there are more x's on the bottom than on the top? Pay it no mind, my friend...watch:

Example: Simplify the following expression into one exponent:

$$\frac{x\, x^3\, x^6}{x^2\, x^8\, x\, x^5} = \frac{x^{10}}{x^{16}} = \frac{1}{x^6}$$ This is a perfectly acceptable answer! However, some professors will require that you have the base in the numerator of the fraction. In that case we remember from our little exercise with all the base 10 numbers that $1/x$ is the same as x^{-1}, so $1/x^6 = x^{-6}$. In my humble opinion, either answer is ok!

Professor's Practice Pause

(1) In your own words define the math term *exponent*.

(2) Give two ways in which exponents are written.

(3) Using 2 as your base and exponents from -3 to +3, write out 2 to each exponent and show how the exponent rules are the same as using 10 for the base (like we did above).

(4) Without the use of your TI, evaluate the following exponential numbers:
(a) 3^{-3} (b) $(-2)^3$ (c) 1548^0 (d) 32^1 (e) 32^{-1}

(5) Simplify the following expressions into just one exponent:

(a) $\dfrac{y^5\, y\, y^2\, y^3}{y^5\, y^0\, y^2} =$

(b) $\dfrac{z^2\, z^{(-3)}\, z}{z^{(-5)}} =$

Associated End of Booklet exercises are questions 1 – 5.

Signs, Numbers, Letters (also called Saturday Night Live)

How many of you have ever watched Saturday Night Live (SNL)? By now they probably have their own channel. I wouldn't know, as I write this I only have 1 TV channel – and that one's fuzzy! In math, SNL stands for **Signs Numbers Letters**.

ACRONYM EXERCISE

Write your own acronym for SNL that crisply summarizes your Academic Advisor.

_____

When we have an expression with all kinds of stuff mixed together, we don't freak out! We chill...then we take it one step at a time – first the signs, then the numbers, and finally one letter (variable) at a time.

Example: Simplify the following expression as much as possible:

$$\frac{-6\,(x\,y\,z)^2\,y\,z^{(-1)}}{2\,x\,(3\,y)^2\,x^3}$$

First we look at the **Signs**: There is a negative on top and no negative on the bottom, so the answer is negative. Whenever the total number of negatives is odd, the answer is <u>negative</u>.

Now the **Numbers**: If we break them out, we have $\dfrac{6}{2\,3^2} = \dfrac{6}{18} = \dfrac{1}{3}$

One **Letter** at a time:

> We really have $z^2 \cdot z^{-1}$
> which $= z^{(2-1)} = z^1 = z$

Counting up all the x's, we have $x^2/x^4 = 1/x^2$ or x^{-2}.
Counting up all the y's, we have $y^3/y^2 = y$.
Counting up all the z's, we have just z on the top.

Now put everything together...

$$\frac{-6\,(x\,y\,z)^2\,y\,z^{(-1)}}{2\,x\,(3\,y)^2\,x^3} = -(1/3)x^{-2}yz, \text{ or } -\frac{y\,z}{3\,x^2} \quad \underline{\textit{Why are these two representations the same?}}$$

Very Important TI-83 note...Group Discovery

Remember, your calculator cannot do symbolic manipulations. It cannot, for example, correctly simplify the example we just did above. Try this little exercise to prove it with the rest of your class. Each of you type in "x-squared times x-cubed." That is x²*x^3 and then press **ENTER**. Notice you <u>do not get</u> the correct answer x⁵, you get a number. And if your Professor will list everyone's numbers on the board, you will notice that they won't even be the same! The reason is that the TI stores a specific value for the number x – you can find out yours by simply typing x and then **ENTER**. The TI is stupid, but very, very fast!

Group Pause

In groups of 2 to 4, use SNL to simplify the expressions below. Work together and help each other. The best way to learn math is to be able to teach math to your math neighbor!!

(a) $\dfrac{(-2x)^2 y^3 \, 4z}{-8 \, x^3 \, z^3 \, y^2}$

Sign:

Number:

Letter x:

Letter y:

Letter z:

Simplified answer =

(b) $\dfrac{(p^2 q)^3 \, (3t)^2}{-q^0 p^0 \, (-2t)}$

(c) $\dfrac{x^{(-3)} \left(\dfrac{y}{2}\right)^{(-2)} z}{\left(\dfrac{z}{3}\right)^4 \, x\,y}$

(d) $\dfrac{\left(2\,x^2\,y\,2\,z^2\right)^{(-1)}}{-3\,x\,y\,z}$

Associated End of Booklet exercise is question 6.

Radicals (or roots)

If exponents are the same as *powers*, then radicals are the same as *roots*. What do you think of when you see this symbol... $\sqrt{}$? Most people know that it means "the square root," but why does it mean that? Certainly you wouldn't just guess that from looking at the symbol itself, right? And then of course you can have "cube roots" and "fourth roots" $\sqrt[3]{}$ and $\sqrt[4]{}$

It is believed by many mathematical historians that the first entrance of this symbol into the mathematical world was in Christoff Rudolff's book *Die Coss*, written in 1525. The reason for the title is that cosa is a thing which was used for the unknown. Algebraists were called *cossists*, and algebra was called the *cossic* art for many years.

As good as Rudolff's work was, it would make a lot more sense if we could express these radicals as exponents because we know how to use exponents. The radical symbol is not mathematically intuitive. Well there is a way, but what exponent is the "square root"?

Algebraically it looks like this...we know that $(\sqrt{x})(\sqrt{x}) = x$ ...try it...$(\sqrt{100})(\sqrt{100}) = 100$

So, let's represent the radical as the letter n

$(x^n)(x^n) = x$...which means that $x^{2n} = x^1$

If the bases are the same that means the exponents have to be the same, so we equate the exponents...

$2n = 1$...and $n = \frac{1}{2}$

The $\sqrt{x} = x^{\frac{1}{2}}$. By a similar argument, the cube root of $x = x^{1/3}$ and, in general, the "nth root of x" (that is, $\sqrt[n]{x}) = x^{1/n}$. So whenever I see something like this ... $\sqrt{(x-5)}$...I will change it to $(x - 5)^{\frac{1}{2}}$ and so on...

The *Radical* form is...	The *Exponent* form is...
$\sqrt{x}$	$x^{1/2}$
$\sqrt[3]{x}$	$x^{1/3}$
$\sqrt[4]{x}$	$x^{1/4}$
$\sqrt[n]{x}$	$x^{1/n}$
$\sqrt[3]{x^2}$	$x^{2/3}$
$\sqrt[n]{x^m}$	$x^{m/n}$

Notice that with both powers and roots, the power is written in the numerator and the root is written in the denominator.

$x^{5/2}$ is the "square root of x to the fifth" or "x to the five-halves power"

$9^{3/2} = (9^{1/2})^3 = (\sqrt{9})^3 = 3^3 = 27$

Some numerical examples:

Without the use of a calculator, let's evaluate these exponential numbers. Change any radical forms to exponential forms first.

(a) $81^{3/2}$ (b) $\sqrt[4]{4^2}$ (c) $\sqrt[3]{x^{(-2)}}$ (d) $(1/16)\wedge(-3/2)$

(a) $81^{3/2}$ means "The square root of 81 then cubed," but it also means "81 cubed, then the square root." It's almost always easier to make the root smaller first, so let's try the first meaning…$81^{1/2} = 9$ and $9^3 = 9*9*9 = 729$

(b) $\sqrt[4]{4^2}$ means "the fourth root of four squared." Changing to exponential form (power on top, root on the bottom), we have $4^{2/4} = 4^{1/2} = 2$.

(c) $\sqrt[3]{x^{(-2)}}$ when changed to exponential form is $x^{-2/3}$…that's it. If we wanted to make the exponents positive we would write $1/x^{2/3}$.

(d) $(1/16)\wedge(-3/2)$. This problem is written in TI-83 (or caret) form. By re-writing with a positive exponent, we have $1/(1/16)^{3/2} = 16^{3/2} = (16^{1/2})^3 = 4^3 = 64$

What do the following radicals represent?

$\sqrt{beer}$ $\sqrt{of\ all\ evil}$

$\sqrt{canal}$ $\sqrt{mean\ square}$

A Method of Approximating Square Roots

Working with square roots and finding algorithms to approximate them quickly and accurately has kept many mathematicians awake to the wee hours of the morning. The ancient Babylonians, Chinese, and Greeks had methods. In Medieval Europe, Leonardo of Pisa and Jordanus de Nemore built upon older methods to come up with their own approximation algorithms. Why even Sir Isaac Newton, after eating many apples, contributed his algorithm for approximating square roots. In the 17th century, the English mathematician John Pell found what he called Pell numbers, sequences of integers that followed certain patterns. ((Note: we will assume this algorithm comes from Pell, although some mathematical historians believe that credit was given to Pell erroneously.)) The algorithm that follows can be used to find the square root of any number…

Warning – big word alert

The algorithm is *recursive*. We need to define a few variables because we will repeat this pattern over and over until we have a close estimate. Let's start by approximating $\sqrt{5}$, which is about 2.236. We will always start with a guess and write that guess as a fraction.

2 is a reasonable guess because $2^2 = 4$, which is close to 5. We write 2 as 2/1. We will call our guess, **G**, the numerator of our guess, **N**, and the denominator of our guess **D**. N_n is the next numerator and N_p is the previous numerator – Likewise for **D**. Since the number 5 is the root, we will call that **R**. R_G represents our guess "squared", which is an approximation for the root. The letter **E** represents our percent *error* from the actual square root. The algorithm looks like this…

ALGORITHM
(1) $N/D = G$
(2) $G^2 = R_G$
(3) $\lvert R_G - R\rvert/R = E$
(4) $N_n = D_p * R + N_p$
(5) $D_n = D_p + N_p$
(6) Go back to step (1)

TRIAL 1

$G = 2, N = 2, D = 1, R = 5$

(1) $2/1 = 2…$2 is our first guess
(2) $2^2 = 4…$4 is our approximation to 5
(3) $\lvert 4 - 5\rvert/5 = 0.2 = E…$our first guess has 20% error
(4) $N_n = 1*5 + 2 = 7$
(5) $D_n = 1 + 2 = 3$ Our next guess is 7/3
Notice our first guess is low
$4 < 5$

TRIAL 2

$G = 7/3, N = 7, D = 3, R = 5$

(1) $7/3 = 2.\overline{3}…$our second approximation for $\sqrt{5}$
(2) $(7/3)^2 = 49/9 \sim 5.\overline{4}…$our approximation to 5
(3) $\lvert 49/9 - 5\rvert/5 = 0.2 = E…$8.8% error
(4) $N_n = 3*5 + 7 = 22$
(5) $D_n = 3 + 7 = 10$ Our next guess is 22/10

Notice our second guess is high
$5.444 > 5$

TRIAL 3

$G = 22/10, N = 22, D = 10, R = 5$

(1) $22/10 = 2.2$
(2) $2.2^2 = 4.84…$new approximation to 5
(3) $\lvert 4.84 - 5\rvert/5 = .032 = E…$3.2% error
(4) $N_n = 10*5 + 22 = 72$
(5) $D_n = 10 + 22 = 32$
Our next guess 72/32

Notice our third guess is low
$4.84 < 5..$this will continue to alternate!

We will stop after 4 trials...

TRIAL 4
G = 72/32, N = 72, D = 32, R = 5
(1) 72/32 = 2.25
(2) $(72/32)^2 = 5.0625$
(3) |5.0625 − 5|/5 = 0.0125 = E...1.25% error

So, after just 4 trials we have an approximation to √5 that is within 1.25% error of the root.

Our approximation is 2.25 and the actual square root is about 2.236

It will only take a few more trials before we approach 0% error.

Look at the sequence of approximations for the $\sqrt{5}$ ...

2/1, 7/3, 22/10, 72/32,... If we keep going, we have

$$2/1, 7/3, 22/10, 72/32, 232/104, 752/336, 2432/1088, ..., \frac{D\,R+N}{D+N}$$

In other words, we always find the next approximation by the root by the denominator, adding the numerator and dividing that by numerator plus denominator. Keep in mind also that $\sqrt{5}$ is an irrational number. All of our successive approximations are rational – which of course means that we will never actually get to the exact square root.

Let's try the same for $\sqrt{2}$ guessing 1 and just using the result, $\frac{D\,R+N}{D+N}$. We know that the approximation generally used is 1.414.

$$1/1, \quad \frac{1\,(2)+1}{1+1} = 3/2, \quad \frac{2\,(2)+3}{2+3} = 7/5, \quad \frac{5\,(2)+7}{5+7} = 17/12, \quad \frac{12\,(2)+17}{12+17} = 41/29,$$

$$\frac{29\,(2)+41}{29+41} = 99/70 = 1.41429 \text{ whereas } \sqrt{2} \text{ is approximately } 1.41421.$$

99/70 is an excellent, rational approximation for $\sqrt{2}$ with only 0.006% error...Now that's close!

Professor's Practice Pause

(1) Without using your TI-83, evaluate the following exponential numbers. Change any radical forms to exponential forms first.

(a) $8^{1/3}$ (b) $27^{2/3}$ (c) $\sqrt[3]{125}$ (d) $(1/64)^{\wedge}(-1/3)$

(e) $\sqrt[3]{x^5}$ (f) $81^{3/4}$

(2) Using our approximation algorithm, find a rational approximation for the $\sqrt{3}$ with less than 1% error.

Associated End of Booklet exercises are questions 7 – 11.

Is it possible to hear $\sqrt{2}$?

We have written $\sqrt{2}$. We have spoken $\sqrt{2}$. We have seen $\sqrt{2}$. We have even approximated $\sqrt{2}$. But can we really hear $\sqrt{2}$. You are probably thinking, 'Ok...now he's finally lost it...flipped his lid...got bats in his belfry...his elevator doesn't go all the way to the top...he's a few sandwiches short of a picnic' and all those crazy things. But can we actually hear the $\sqrt{2}$. As the great, fictional detective of the British moors, Sherlock Holmes, would say, 'Elementary, my dear Watson! Elementary...'

In order to do this, we have to go back to the Western Chromatic Scale from the Booklet on Real Numbers, but then jump forward to something you haven't learned yet in the Booklet on Exponential Functions...

Future Memory Break

We did not learn yet that in order to get from the frequency of one note in the scale to the frequency of the next note in the scale (one half-step up) we need to multiply the frequency of the first note by $2^{1/12}$. For example, in the table below, the frequency of the note C is 1000Hz. In order to calculate the frequency of the note C# (the next note in the scale), we would find $1000(2^{1/12}) = 1059.46$Hz.

Note	Pure Frequency
C	1000.00Hz
C#	1059.46Hz
D	1122.46Hz
D#	1189.21Hz
E	1259.92Hz
F	1334.84Hz
F#	1414.21Hz
G	1498.31Hz
G#	1587.40Hz
A	1681.79Hz
A#	1781.89Hz
B	1887.75Hz
C	2000.00Hz

In music theory, in order to go from playing the *root* note to playing the *diminished fifth* note, you would need to go up the scale 6 half-steps. For example, if your root note is C (at 1000Hz), the diminished fifth note is F# (at 1414.21Hz). When these notes are played together, the ratio of the frequencies is equal to $\sqrt{2}$.

We multiply 1000 by $2^{1/12}$ six different times. But we know from our rules that $(2^{1/12})(2^{1/12})(2^{1/12})(2^{1/12})(2^{1/12})(2^{1/12}) = 2^{6/12} = 2^{1/2} = \sqrt{2}$. Notice also that the ratio of the frequencies themselves is $1414.21/1000 = 1.41421$ which, from before, is our approximation for $\sqrt{2}$.

Notice that in going from one note to the same note an *octave* higher, we double the frequency. This is because we would be multiplying by $2^{1/12}$ twelve separate times, or $2^{12/12}$, which is 2^1, or simply 2 – hence the doubling.

The Laws of Mathematics

There are many laws that are used in math, but I want us to focus on the ones that will be used most often. And if you remember correctly, I've said several times that I don't want you to memorize rules, but rather learn to derive principles instead. Hopefully, I can show you this with the <u>Distributive</u>, <u>Commutative</u>, and <u>Associative</u> laws. Think *distributing*, *commuting*, and *associating...*

The Distributive Law

You use the distributive law all the time in conversation and you probably don't even realize it. If I told you to <u>buy more soda and chips</u>, what would you buy?...correct, more soda and more chips. So,

more (soda and chips) becomes more soda and more chips.

((watch as we slowly transform...))

m(s and **c)**...becomes **ms** and **mc**

((*and* means +...so))

m(s + c) = **ms** + **mc**

Viola! What we just saw was an example of the Distributive Property in math played out in English. You are *distributing what is outside the parentheses to everything that is inside the parentheses.*

Likewise, "5 times 2 and x", or 5(2 + x), is really "5 times 2 and 5 times x", or 5(2) + 5(x).

<u>Examples</u>:
    (a) $-3(x + y) = -3(x) + -3(y) = -3x - 3y$
    (b) $-3(x - y) = -3(x) - -3(y) = -3x + 3y$
    (c) $3(x + y - z) = 3x + 3y - 3z$
    (d) $A(B - C) = AB - AC$

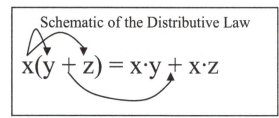

Schematic of the Distributive Law

$$x(y + z) = x \cdot y + x \cdot z$$

The Commutative Law

What does it mean to be a *commuter*? A person who travels to work back and forth. Commuting is going back and forth. In math, when something commutes, it works the same one way as it does the other way. The easiest example of that is multiplication...we all know that $2 \cdot 5 = 5 \cdot 2$. They both equal 10. Five groups of 2 is the same as 2 groups of 5, and so on. Another way to say that is ab = ba.

What other math operation works the same one way as the other way? Addition. $2 + 5 = 5 + 2 = 7$. Another way to say that is $a + b = b + a$

What about subtraction and division? They do not commute (for some people they don't compute either!!). $2 - 5 \neq 5 - 2$. The answers are -3 and 3. They are not the same, but they are related. Notice that they have the same value but different signs. How about division? $2/5 \neq 5/2$. One answer is 0.4 and the other is 2.5. Unlike subtraction, they will have the same sign but different values.

Multiplication and addition are like taking the train to work and then taking the train back home. Subtraction and division are like taking the train to work and then catching a taxi to the airport, flying to Rome, and hanging out on the beach for a week!

Examples:
> (a) $3 + 11 + 6 = 6 + 11 + 3 = 20$
> (b) $5 \cdot 6 \cdot 2 = 2 \cdot 5 \cdot 6$
> (c) $ABC = CBA$
> (d) $A + B + C = C + B + A$

The Associative Law with a twist!

You're all grown-ups now so you can *associate* with whomever you wish. That's what the associative law in math says. Instead of writing $A - B - C + D$, I can choose to associate the positives and associate the negatives. Let the positive's chill with the positives and the negatives marinate with the negatives. So I can re-write that as $A + D - B - C$. As long as we keep the original sign with the proper term, we can switch them around (associate them) however we want.

> **Example**: Evaluate $-3 + 5 - 4 - 2 + 7 + 6 - 8$. If we choose, we could simply mover from left to right and add and subtract as we go. but, we could also associate +'s and –'s together, like so...$5 + 7 + 6 - 3 - 4 - 2 - 8 = 18 - 17 = 1$.

$$+18 \qquad -17$$

Examples:
> (a) $AB(CD) = (AB)CD = A(BC)D$
> (b) $A + (B + C) = (A + B) + C$
> (c) $8 - (4 - 1) \neq (8 - 4) - 1$ because the left side is $8 - 3 = 5$, but the right side is $4 - 1 = 3$...this is a counter-example.

Let's put all 3 together and see how we can use them to make math a bit easier...

Simplify 3(-5 – x + 2 + 4x)

The Distributive law says…3(-5) + 3(-x) + 3(2) + 3(4x) = -15 – 3x + 6 + 12x
The Associative law says…-15 – 3x + 6 + 12x can be written as (-15 + 6) + (-3x + 12x) = -9 + 9x
The Commutative law says…-9 + 9x can be written as 9x – 9.

When we get to factoring polynomials in the Booklet on Quadratic Functions we'll see that we can turn 9x – 9 into 9(x – 1)

Prime Factorization

Lettuce remember what a prime steak number…I'm sorry – I'm getting hungry! Going back to prime numbers from the Booklet on Real Numbers, we had this definition…

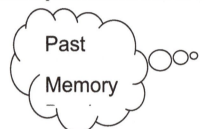

A *prime* number is a positive integer whose only divisors are 1 and itself. A *composite* number is not prime – it is divisible by some other number.

We can factor composite numbers using something called a Prime Factorization Tree. It looks something like this…

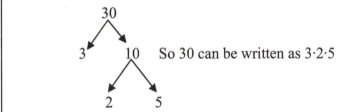

So 30 can be written as 3·2·5

The trick is to continue breaking down the branches until only prime numbers remain, then write the original number as the product of all the prime leaves (which are the prime numbers)

Example: Factor 180 completely into primes

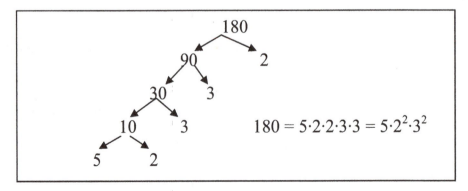

$$180 = 5 \cdot 2 \cdot 2 \cdot 3 \cdot 3 = 5 \cdot 2^2 \cdot 3^2$$

Here's a for-true…every composite number can be written as a product of prime numbers in one and only one way!! Now that's something you can take to the bank. I bet Mrs. Bumblemeyer never told you that, did she?! $180 = 5 \cdot 2^2 \cdot 3^2$ …that's it! It cannot be written in any other way.

We will use these skills when factoring polynomials, but they also help build that math muscle we were talking about earlier. Here's another example…

Example: Factor 17640 completely into primes.

Again since the number is even we know we can always cut it in half…

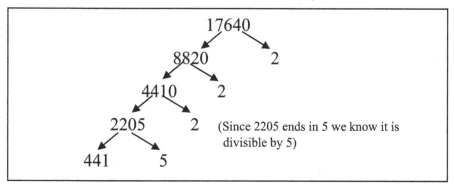

    (Since 2205 ends in 5 we know it is divisible by 5)

We would seem to be stuck now, however here's another fascinating property of numbers which we you may have heard at some point in your mathematical careers…. If the sum of the digits of a number add up to 9 (or a multiple of 9), the number is divisible by 9 – which also means it is divisible by 3 twice!

    441 / 3 = 147. 147 / 3 = 49. And, of course, 49 = 7*7

So, 17640 can be uniquely written as $2^3 * 3^2 * 5 * 7^2$.

Group Pause

In groups of 2 or 3, use the space below to factor 207900

Associated End of Booklet exercises are questions 12 – 22.

The Order of Operations (Please Egg My Dad And Sister)

A very important part of any language is its grammatical structure. This is true whether we are talking about spoken languages around the world, computer programming languages, or the language of mathematics.

Languages can have different grammatical orders. For example, English tends to be a language that follows the order of noun, verb, object. Some languages, like German, more frequently use inverted word order: noun, object, verb.

Languages can be largely *idiomatic*, in which the meaning of phrases cannot be deduced from the literal translation. For example, the German phrase *Ich habe Hunger* and the Italian phrase *ho fame* mean "I have hunger", but in English we would say *I am hungry*, a more literal translation.

Some languages are read from left to right, while others are read from right to left. Some Asian languages are read from top to bottom.

There are even languages that are *metaphorical*, such as the language of the Children of Tamar, in episode 502 of *Star Trek: The Next Generation*. The language is deeply rooted in local historical events. For example, "Darmok and Jalad at Tanagra," refers to the situation in which heroes must learn to trust each other so that they may work together to defeat a common foe.

As a for-true language, math has its own grammatical structure as well. We call this structure, the order of operations (OOO, for short or O^3 for even shorter). The OOO allows everyone on the planet (Earth) to "speak" math in the same fashion and arrive at the same results regardless of where the problem is done or who does it.

The order in which operations are done in math can be broken down to 4 levels: Parentheses, Exponents, Multiplication/Division, and finally Addition/Subtraction. If you consider only the first letters of each of those words, you come up with PEMDAS.

"**P**lease **E**gg **M**y **D**ad **A**nd **S**ister," was on the note found during the Easter egg hunt.

Or, maybe you remember…

"**P**lease **E**xcuse **M**y **D**ear **A**unt **S**ally," said the young man, "She doesn't do math very well!"

ACRONYM EXERCISE
Write your own acronym for PEMDAS to help you remember the OOO

_____

For example, look at the expression below:

$$4(-3 - 7)^2 + 5/3 - (16 + 9)/3^3$$

If everyone just picked their own individual way to simplify the expression above, everyone would come up with a different answer and we wouldn't be able to communicate mathematically.

Level 1: Parentheses. We do as much as we can inside parentheses, or brackets "[ ]", or squigglies "{ }". Parentheses are done when you can't simplify what's inside any more. When dealing with only numbers, the parentheses can then be removed.

$$4(-10)^2 + 5/3 - (25)/3^3$$

Level 2: Exponents...$(-10)^2$ is $(-10)(-10) = 100$ and 3^3 is $(3)(3)(3) = 27$, so that gives us :

$$4(100) + 5/3 - (25)/27$$

Level 3: Multiplication and Division from left to right. If we change our answers to decimals as we go we would have:

$$400 + 1.67 - 0.926$$

Level 4: Addition and subtraction which we also do from left to right (even though it doesn't really matter, right?...Associative Law).

$$401.67 - 0.926 = 400.744$$

Another note about decimals vs. fractions. It doesn't matter which one you use or prefer just as long as you are consistent. If you prefer fractions then keep everything in the problem as fractions, and likewise with decimals.

In this next example, I am going to write the same problem in **Textbook** math and **TI-83** math, two different dialects of the math language.

Example: Simplify the following:

$$\frac{\left((5 - 2^2)^2 - \frac{-4 - 6}{2}\right)^3}{10}$$ ...or... $((5 - 2\char`^2)\char`^2 - (-4 - 6)/2)\char`^3/10$

I will use the TI-83 dialect to simplify this problem...Notice that within each set of parentheses, you may have to use some or all of the other 3 Levels before saying that set of parentheses is complete.

Level 1: $((5 - 2^2)^2 - (-4 - 6)/2)^3/10$....working on the 2 inner sets of ( )
      E: 5 – 4   S: -10
      S: 1

Re-write: $(1^2 - -10/2)^3/10$....now working on the outer set of ( )
       E: 1   D: -5

Re-write: $(1 - -5)^3/10$....finish simplifying the outer set of ( )
       A: 6

Re-write: $6^3/10$........we just finished all Level 1's (P), now we move to Level 2

Level 2: 216/10

Level 3: 21.6........Answer.

Example: In the Booklet on Exponential Functions, we will study different types of interest that your money can earn. *Simple* interest is calculated using the formula $P(1 + i)^t$, where P is the amount invested (Principal), i is the interest rate (in decimal), and t is the number of years.

      Find the total amount of money if $1000 is invested at 5% for 2 years.
      Amount $= 1000(1 + .05)^2$

Level 1: $1000(1 + .05)^2$
        A: 1.05

Re-write: $1000(1.05)^2$

Level 2: $(1.05)^2 = (1.05)(1.05) = 1.1025$

Re-write: 1000(1.1025)

Level 3: 1102.5......Answer $1102.50

Example: Simplify the following expression... $\dfrac{-13\,(5-7)^2 - 6}{10} + 5^3$

$\dfrac{-13\,(-2)^2 - 6}{10} + 5^3$ ...simplify the parentheses

$\dfrac{-13\,(4) - 6}{10} + 125$ ...simplify the two exponents...$(-2)^2 = 4$ and $5^3 = 125$

$$\frac{-52 - 6}{10} + 125 \qquad \ldots\text{multiplication}\ldots -13(4) = -52$$

We cannot divide by 10 until we simplify the top of the fraction. Essentially -52 – 6 can be written using parentheses as (-52 – 6), so we want to go back and complete the parentheses first.

$$-\frac{58}{10} + 125 \quad \ldots\text{simplify the numerator}\ldots\text{now we can divide}\ldots$$

-5.8 + 125 = 119.2……Answer

> *Nested* means parentheses within parentheses within parentheses…

Example: Simplify the following expression: $1 - (1 - (1 - (1 - 2)))$

HINT: When you have *nested* parentheses, always work from the inside out. If this book was in color, you would have seen exquisite uses of color to distinguish different sets of parentheses.

$1 - (1 - (1 - (-1)))\ldots..1 - 2 = -1$

$1 - (1 - (2))\ldots\ldots 1 - (-1) = 1 + 1 = 2$

$1 - (-1)\ldots..1 - (2) = -1$

$= 2\ldots\ldots 1 - (-1) = 2$

Other random things you should know…

These are some things that you may never use and if you needed to you could probably just look them up somewhere, but we think they are worth mentioning. **F**irst, every budding young mathemagician should know the Greek alphabet (In fact, the word *alphabet* is a <u>concatenation</u> of the first two Greek letters, alpha and beta)

> **Warning** – big word alert

alpha	α*	zeta	ζ	lamda	λ	pi	π	phi	φ
beta	β	eta	η	mu	μ	rho	ρ	chi	χ
gamma	γ	theta	θ	nu	ν	sigma	σ	psi	ψ
delta	δ	iota	ι	xi	ξ	tau	τ	omega	ω
epsilon	ε	kappa	κ	omicron	o	upsilon	υ		

* These are the lower case letters.

Remember that it was the Greeks who developed almost all of modern day Euclidean geometry – you can thank Euclid for all the geometrical proofs that you did in high school. We honor the Greek mathematicians by using just about every letter in their alphabet for math, science, statistics, biology, engineering, you name it! You have probably seen some of these letters already. The dreaded angle in trigonometry is almost always either *theta* or *phi*. If you have learned about standard deviation in statistics, you know that we use *sigma*. The letter *chi* is used quite often in thermodynamics. Our symbol for "micro" is the letter *mu*. In the Bible, in the book of the Revelation, Jesus calls himself the *alpha* and *omega* (the beginning and the end).

Like Hebrew, Greek letters are also used to count. Alpha is 1, beta is 2, and so on. Sounds funny, doesn't it? Remember Roman numerals? Could you imagine, lining up at the cashier at Wal-Mart, 'uh, that'll be *cguzy* dollars and *iiv* cents please'..?!?!

Secondly-ish, since our country is one of three countries in the world that does not use the Metric System (**can you name the other 2**?) we thought it would be important to review the Metric System prefixes once again. Mrs. Bumblemeyer probably taught them to you as well…

Prefix	Numerical Meaning	Example	Symbol
Tera	10^{12}	Tera floating point operations per second	Teraflops
Giga	10^{9}	Gigabits per second	Gbps
Mega	10^{6}	Megawatts	Mw
kilo	10^{3}	kilometer	km
hecto	10^{2}	hectoliter	hl
deka	10	dekagram	dag
deci	10^{-1}	deciliter	dl
centi	10^{-2}	centimeter	cm
milli	10^{-3}	milliamp	ma
micro	10^{-6}	microvolt	μv
nano	10^{-9}	nanometer	nm
pico	10^{-12}	picogram	pg

Of course, having an understanding of the Metric System implies (and incorrectly so) that one has an understanding of…Scientific Notation – **T**hirdly. Again, the not altogether unstoppable Mrs. B surely taught you scientific notation, however since you *rarely* used a calculator in middle school (ha! ha! ha!), you may have forgotten.

Intended to be read with dripping sarcasm…

In the natural science and engineering, we deal with….I say *we* because I used to be an engineer…we deal with very, very large and very, very small quantities. The number of miles from the Earth to the nearest black hole, the length of the nucleus of an atom in meters, the number of molecules in a certain volume, and on and on…By the way, the 12-step program established for people who can't stop talking is called "on and on anon". So, instead of writing a

** The other two are Liberia and Myanmar

million 0's after a number we have created a compact notation for playing with these big numbers. For example:

(a) *Avogadro's* constant is the number of molecules in a mole and is approximately 60221417900000000000000. If you've taken chemistry, you know that you write this number quite a bit, right? So there's really no reason to do that all the time when we can simplify by doing the following:

60221417900000000000000.0	move the decimal to just after the 6 and count how many places the decimal was moved
6.0221417900000000000000	23 places, and each place is a multiple of 10
6.022×10^{23}	drop the rest of the numbers and add the number of decimal moves as an exponent on the number 10.

We read this as, "Six point oh two two times ten to the twenty-third molecules"

(b) The weight of the Earth is 5.972 sextillion metric tons, which is 5972000000000000000000 metric tons. And since a metric ton is 1000 kg, that's about 5972000000000000000000000 kg. And since 1 kg is 1000g, that's about 5972000000000000000000000000 g. Again, we wouldn't want to keep writing this number with all the 0's, so we do the same 3 steps we did above…

5972000000000000000000000000.0	move the decimal to just after the 5 and count how many places the decimal was moved
5. 972000000000000000000000000	27 places, and each place is a multiple of 10
5.972×10^{27}	drop the rest of the numbers and add the number of decimal moves as an exponent on the number 10.

The Earth weighs "five point nine seven two times ten to the twenty-seventh grams"

(c) The mass of an electron is about 0.00000000000000000000000000000091094 Kg. We don't want to keep writing this really, really small number so we do the same three steps, with one slight difference. Since the decimal place is moving right, our exponent is *negative*.

Moving the decimal 31 places to the right brings us to 9.1, so the mass can be represented as 9.1094×10^{-31} Kg.

Working with Scientific notation follows the properties of exponents that we have used before. Imagine this: What would you say if I told you to multiply 3245000000000 by 90100000? You would say, like, later dude!

But, if we use Science fiction Notation and change those numbers to 3.245×10^{12} and 9.01×10^7, it becomes much easier. We multiply the number parts and add the exponents together because they both have base 10...

$(3.245 \times 10^{12}) (9.01 \times 10^7) = (3.245)(9.01) \times 10^{(12 + 7)} = 29.237 \times 10^{19}$. And, if we want to represent our answer with no tens place we could write 2.9237×10^{20} (because we moved the decimal one place to the left)

Example: Scientists believe that the number of electrons existing in the known universe is at least 10^{79}. This number amounts to an average density of about one electron per cubic meter of space. Of course that's only because space is mostly empty - ...I was about to say... empty *space*...duh! Given the mass of an electron as 9.1094×10^{-31} Kg, find the combined mass of all the electrons in the universe.

$$\text{mass of all electrons} = (\text{number of electrons}) (\text{mass of 1 electron})$$
$$= (1 \times 10^{79}) (9.1094 \times 10^{-31})$$
$$= 9.1094 \times 10^{(79-31)} = 9.1094 \times 10^{48} \text{ Kg.}$$

Since there are 1×10^3 grams in 1 Kg, the mass of all the electrons in grams would be:
$$= (9.1094 \times 10^{48}) (1 \times 10^3)$$
$$= 9.1094 \times 10^{51} \text{ g}$$

By comparison, the mass of our sun is about 2×10^{33}g. If we divide the mass of all the electrons by the mass of our sun we get...

$$\frac{\text{mass of all electrons}}{\text{mass of our sun}} = \frac{9.1094 \times 10^{51}}{2 \times 10^{33}} = (9.1094/2) \times 10^{(51-33)} = 4.55 \times 10^{18}$$

The mass of all the electrons in the universe is about the same as the combined mass of 4,550,000,000,000,000,000 of our suns. That's 4.55 *quintillion* suns!

Bill Nye the Science Guy's got nothin' on us, yo!!

Example: In 2009, the National Debt is approximately $10 trillion (that's 13 zeros). The number of people in the US is about 300 million. On average, how much does each person owe?

Average debt of 1 American is the National Debt divided by the number of Americans.

Average Debt = $1 \times 10^{13} / 3 \times 10^8 = .333 \times 10^{(13 - 8)} = .333 \times 10^5 = 3.33 \times 10^4 = \$33,300$.

If we each paid about $33,000, we would no longer have a National Debt....I report, you decide (just like FOX news).

Professor's Practice Pause

For problems α to δ, simplify the following expressions using the Order Of Operations.

α) $\dfrac{|-25|-2\,(-5)}{2^4-9}$

β) $-3[5^2-(7-3)^2]$

γ) $27^{2/3}-8(3!)^2/(-1-2^3)$

δ) $(4^2+3)^2-5-\dfrac{\dfrac{1}{2}+\dfrac{2}{3}}{6}$

ε) A capacitor is an electronic device that stores electrical energy. The unit of measurement for capacitors is called the *farad*. Use scientific notation to express a 478 picofarad capacitor to 2 decimal places.

ζ) Given that $A = 3.04 \times 10^{-29}$ and $B = 2.21 \times 10^{46}$, find:
(1) A·B   (2) A/B

Associated End of Booklet exercises are questions 23 – 25.

TI - 83 Basics

I. The first thing you notice about the TI-83 is the different colors on the face – black, yellow and green. The main function of each key is what you see in black. If you want to get to the yellow function of the key you must first press the **2ⁿᵈ** key. For example, to access **QUIT**, you must press **2ⁿᵈ MODE**. If you want to access a green function you must first press the **ALPHA** key. For example, to get the letter "P" you would press **ALPHA 8**.

II. There is no "=" key. To end an operation, press **ENTER**. For example, to find 3 + 5, you would press **3, +, 5,** and then **ENTER** – as below…

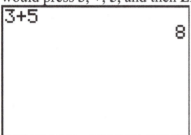

```
VERY USEFUL TIP…

To get back to the home screen, press the
CLEAR button or 2ⁿᵈ MODE (which is
QUIT)
```

III. To set the number of decimal places press **MODE**, arrow down to Float, and then arrow over to the number you want (3 is probably best) and press **ENTER**.

IV. To change an answer back and forth between decimal and fraction press **MATH** after doing the calculation and press either **1:** or **2:**. As an example, I will change 7/5 from decimal to fraction and 1.245 to a fraction.

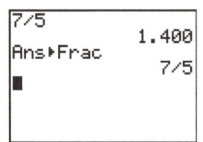

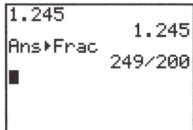

Notice in the last pane on the right that π could not be turned into a fraction…does that surprise you? I hope not…why not?.....**Write your answer over here** ⟶

V. To find the absolute value of a number, press **MATH**, arrow over to **NUM** and press **1:** or **ENTER**. type in the number, close the parentheses) and press **ENTER**.

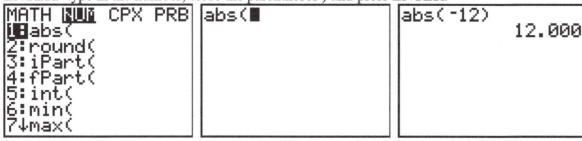

VI. To find the factorial of a number, enter the number on the home screen and then go to **MATH**, arrow over to **PRB**, and press **4:** then press **ENTER** at the home screen.

10! is muy grande

VII. What's the difference between $(-2)^2$ and -2^2? Let's see…

(-2)²	4.000
-2²	-4.000

$(-2)^2 = (-2)(-2) = (-)(-)(2)(2) = 4$

but

$-2^2 = -1 \cdot 2^2$ (the negative of 2^2) = -4

VIII. The order of operations (otherwise known as PEMDAS, otherly otherwise known as OOO, and certainly heretofore previously otherwise known as O^3)

The TI-83 (and 83 Plus and 84 and 84 plus, etc.…) does order of operations implicitly. It is a good skill to have to be able to translate textbook math to TI-83 math, so you can use your calculator for more difficult problems.

For example, let's change $\dfrac{\left(7-5+(3-6)^2\right)^3}{\left(4^3-(8.65-2)\right)^2}$ to TI-83 math…

In TI-speak, we would write $(7-5+(3-6)\text{^}2)\text{^}3/(4\text{^}3-(8.65-2))\text{^}2$…and find that it equals:

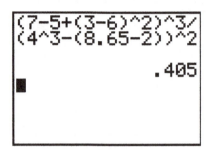

Topical Summary of Basics of the Math Language

Take about 20 – 30 minutes and create your own summary of this chapter. Go back, review, and write below all of the main points, concepts, equations, relationships, etc. This will help lock the concepts in your brain now, and provide an excellent study guide for any assessments later.

End of Booklet Exercises (this, too, shall pass!)

1) What is the definition of "exponent"? What does it tell you to do mathematically?

2) Write 3 different ways of expressing "x cubed"

3) Calculate the following without using a calculator:

(a) 5^2 　　　　　　(b) 2^{-3} 　　　　　　(c) $(-7)^0$ 　　　　　　(d) $(-2)^4$

(e) $(1/2)^{-4}$ 　　　　　(f) 0^2 　　　　　　(g) -2^4 　　　　　　(h) 25^1

4) Use the rules of exponents to simplify the following expressions so that each answer has only one number and one exponent.

(a) $\dfrac{2^2\,2^4}{2^0}$ 　　　　　　(b) $\dfrac{3^3}{3^{(-3)}}$ 　　　　　　(c) $\dfrac{5\,5^{(-1)}}{5^0}$

(d) $\dfrac{4^4\,2^7\,4^{(-2)}}{2^3\,4^0\,2^4}$ 　　　　　　(e) $\dfrac{5^2\,3^4}{5^6}$

5) Use the rules of exponents to simplify the following expressions so that each variable has only 1 exponent. Leaving your answer with negative exponents is fine.

(a) $\dfrac{x^4}{x^7}$ 　　　　(b) $\dfrac{x^7}{x^4}$ 　　　　(c) $\dfrac{x^2\,x^3}{x^4\,x^2}$ 　　　　(d) $\dfrac{z^{(-1)}\,z^2}{z^4}$

(e) $\dfrac{y^0 \, y^{(-2)}}{y^{(-3)} \, y^2}$

(f) $\dfrac{x^n}{x^{(n-1)}}$

(g) $\dfrac{Q^{(-1)}}{Q^{(-2)}}$

6) Use the SNL concept (Signs, Numbers, Letter) to simplify the following expressions as much as possible. Leaving your answer with negative exponents is fine.

(a) $(-3 \, x^2 \, y^4)(-2.5 \, z^3 \, x) \, 10 \, y^2 \, z$

(b) $6 \, x^4 \, y^2 \, (-2 \, y^4 \, z) \, 4 \, z^2 \, x$

(c) $\dfrac{(-3 \, x^2 \, y^4) \, 10 \, y^2 \, z \, (-2.5 \, z^3 \, x)}{6 \, x^4 \, y^2 \, (-2 \, y^4 \, z) \, 4 \, z^2 \, x}$

(d) $\dfrac{5 \, r^3 \, s^3 \, t \, (-2 \, t^2 \, s^{(-1)})}{-10 \, r^2 \, s \, t^{(-2)}}$

(e) $\dfrac{8 \, z^2 \, (z^4 - z)}{32 \, z}$

(f) $\dfrac{((-x)(-y))^2}{(-x)^2 \, (-y)^2}$

(g) $\left(\left(\left(\dfrac{3 \, A \, B^2}{6 \, A^2 \, B} \right)^2 \right)^3 \right)^0$

7) Use the same argument on page 42 to verify that $^3\sqrt{x}$ (the cube root of x) is the same as $x^{1/3}$.

8) Rewrite the following radical expressions as exponents.

(a) $^4\sqrt{y}$

(b) $^4\sqrt{(xy)}$

(c) $^3\sqrt{y^2}$

(d) $^5\sqrt{x^3}$

(e) $^3\sqrt{(x^4y^2)}$

(f) $(\sqrt{y}\ \sqrt{x}\ \sqrt{z})^{\wedge}3$

9) Rewrite the following exponent expressions as radicals.

(a) $x^{\left(\frac{1}{3}\right)}$

(b) $\left(\dfrac{y}{x}\right)^{\left(\frac{1}{2}\right)}$

(c) $\left(x^{\left(\frac{1}{3}\right)}\right)^5$

(d) $y^{\left(-\frac{1}{2}\right)}$

(e) $x^{\left(\frac{2}{3}\right)}$

(f) $y^{\left(-\frac{1}{2}\right)}x^{\left(\frac{1}{2}\right)}$

10) Evaluate the following expressions. Leave your answers in the most exact form.

(a) $2\cdot2^{-1/2}$

(b) $3^{2/3}$

(c) $\dfrac{4^{\left(\frac{1}{2}\right)}}{2^2}$

(d) $^3\sqrt{125}$

(e) $^4\sqrt{81\cdot3^{-1}}$

(f) $\dfrac{\sqrt{\dfrac{1}{2}}}{\sqrt{4}}$

11) Using our approximation algorithm, find a rational approximation for the $\sqrt{7}$ given 5 iterations and a starting value of 3. What is the percent error in our approximation given 2.645751 as the accepted value.

12) In your own words, define the distributive law and give an example using the letters x, y, and z.

13) In your own words, define the commutative law and give an example using the letters x, y, and z.

14) In your own words, define the associative law and give an example using the letters x, y, and z.

For questions 15 – 17, state whether the given sentence is true or false. If it is true, explain why. If it is false, give a counter-example.

15) Multiplication is commutative.

16) Subtraction is commutative.

17) Addition is associative.

For questions18 – 22, find the prime factorization of each number.

18) 35   19) 72   20) 138   21) 1056   22) 47

23) Simplify the following expressions using the order of operations (PEMDAS). If necessary, use 2 decimal places in your answer.

(a) $3((5^2 - 2) + 1)^2$

(b) $\dfrac{\mid -10 \mid - 2^3 + 5}{7\,(3^2 - 2!)}$

(c) .46(2.4 – 6.78)/3 – (5/2)^2 + 1.25

(d) 3(3 – (1 – (2 + (5 – 7))))

(e) $\left(\dfrac{3^0 - 0^3}{4^2 - 2^4} \right)^{\left(\frac{1}{2}\right)}$

(f) $\left(\dfrac{1}{3} + \dfrac{2}{5} - \left(\dfrac{4}{3} - \dfrac{5}{9} \right)^2 - 2 \right)^3$

(g) $|3 - |-7|| + 1$

(h) $\left(\dfrac{2^2 - 1024^{\left(\frac{1}{5}\right)}}{7 + 3^3} \right)!$

24) Rewrite the following numbers using scientific notation. Use the form $O.TH \times 10^n$

(a) .00005

(b) 9,182,000

(c) -32.875

25) Given $X = -5.67 \times 10^{-15}$ and $Y = 1.13 \times 10^{24}$, find:

(a) XY

(b) Y/X

(c) X^2Y^2

Booklet on Complex Numbers

On Complex Numbers, the Worm Hole, and Doppelgänger Theory

The story of complex numbers is a sad one, for it is also the story of our own history – humanity. A long time ago, before any of us were born, all numbers existed harmoniously in a very, well…numerical universe…negatives and positives were friends, rationals and irrationals hung out together…you get the point!

Like everything else however, prejudice crept in. Fear of differences and an unfamiliarity with the potential of numerical diversity raised "10"-sions higher and higher. Who knows how long it took for sure – maybe it was weeks, maybe years. However long, one day all numbers everywhere awoke to separate numerical universes. We call them "Number systems." The Real numbers and the Imaginary numbers were divided by an ocean of space so vast that they had lost all contact with one another. To quote our good friend J.R.R. Tolkien, "History turned to legend and legend to myth."

In the real number system, numbers like 12, 0, -7, and 198.67 were all doing their part in the collective numerical machine….living their daily lives without a thought of 'something else, somewhere – out….*there*."

On the other side of the galaxy, *complex numbers* were doing their thing – equally unaware of the existence of close-kin neighbors. Numbers like $3i$, $5 - 6i$, $12 + i$ did their thing (whatever that thing was??)

You might be thinking, "I've never seen numbers like this!"...and… "What's with all the i's anyway?"

Anatomy of a complex number
(be careful – this is not for young kids or even semi-squeamish students)

You have probably been told sometime in your short mathematical career that you can't have "negatives under the radical," right?? Your High School math teacher (the indomitable Mrs. Bumblemeyer) told you that the following is not allowed…

$$\sqrt{-x}$$

Well… it is allowed – it's just not a real number (those that we deal with on a regular basis). It's an *imaginary* number (a subset of the complex numbers).

In math, we make the following definition for the imaginary unit, i…

$$i = \sqrt{-1}$$

I know, I know…right now you're thinking that the little i thingy stands for imaginary, right? Well, it doesn't. At least I don't think it does. You see, history credits Rene Descartes with coining the phrase "imaginary number," although he said so in a derogatory manner. Leonhard Euler, the great Swiss mathematician furthered the study of complex numbers with the very famous *Euler Relation*. But he spoke French, so he probably didn't use the word "imaginary"…(cause that's English). [For some historical information on Euler, go to Wikipedia.org and type in "Leonhard Euler"].

So this definition tells me that $\sqrt{-5} = \sqrt{5}\ i$

How do I know this? Because our rules of exponents (remember them?) say that $(a\,b)^n = a^n\,b^n$ And since a radical is an exponent, we have the following little mathematical derivation…

$$\sqrt{-5} = \sqrt{5\,(-1)} \quad \text{which equals} \quad \sqrt{5}\,\sqrt{-1}$$

But, since $i = \sqrt{-1}$, we have $\sqrt{-5} = \sqrt{5}\ i$

So then $\sqrt{-4} = \sqrt{4}\ i = 2i$

Furthermore $\sqrt{-25} = \sqrt{25}\ i = 5i$

> In general, $\sqrt{-x} = \sqrt{x}\ i$

Complex numbers are Mutants !! Yes, it's true. What I have showed you up above is the mathematics of *imaginary* numbers. But complex numbers are made up of one part imaginary number and one part real number. Just like *Wolverine* is made up of one part human and one part Adamantium.

| Warning – big word alert |

In fact, (unlike *Wolverine*) complex numbers <u>subsume</u> real numbers. In other words, real numbers exist as a subset of complex numbers. Do you remember the Booklet on real numbers? (yeah, like, uh 6 weeks ago, dude !!) I bring to your memory the relationship between the real numbers…

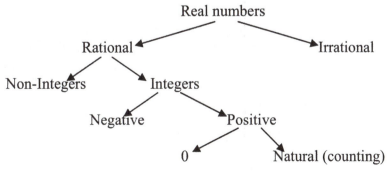

Above the real numbers, we have the following addition to make...

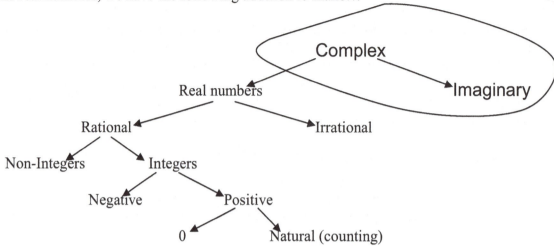

Complex numbers are written as **a + b*i***, where **a** is the real part and **b*i*** is the imaginary part.

{{The following example break is sponsored by *imagina*tion...we think when you don't want to!!}}

Examples of complex numbers...

$2 + 3i$ $-6 + 17i$ $1 - 256i$

$-2 + 0.75i$ $14 + 14i$

$0 + i$ (is this really a complex number? Why or why not?)

$7 + 0i$ (is this really a complex number? Why or why not?)

You get the idea. I know, "What happened to the story?"...ok...ok, already!

Our Story Continues...The Worm Hole...

If any of you have been through an airport recently, you know how tight the security is. Everything gets checked...even your shoes! Some airports have tighter security than others. Newark, NJ is a tough one and so is JFK down in NYC. But there is no security system tighter than the one between the Real and Complex number systems. You see – there is a "Worm Hole" that allows very quick travel between the 2 systems.

For all of you that are not die-hard trekkies, a worm hole is disturbance in space-time that allows almost instantaneous travel between two points that are light years away. Customs and Security at the entrance to the Real number system is impassable – they have light rays and photon cannons, and all that stuff that doesn't really exist.

However, complex numbers have been evading this system and entering the Real world unnoticed. Of course, they would never admit to that, but it's true…I've actually seen some of them.

It turns out that the Worm Hole is nothing other than the **Magnitude** (or *absolute value*) of a number. Complex numbers have been turning themselves into Real numbers using their magnitude for at least a few weeks now. They come over for shopping, maybe visiting some old friends and neighbors from before *The Separation*…who knows why?

So, do you want to know how they do it? To see this, we need to take a deeper look at the world of complex numbers – where they live, what they eat, etc. In an attempt that ultimately cost his life, our numerical reporter on the ground gave us this information about…

The Complex Plane
(Our old friend Tattoo, from Fantasy Island, would call it "De Complex plane…de Complex plane")

The Complex plane looks like the x-y coordinate system (for graphing real numbers), with one small difference. Since complex numbers have both real and imaginary parts, the x-axis stays the real axis, but the y-axis becomes the *imaginary* axis….like so….

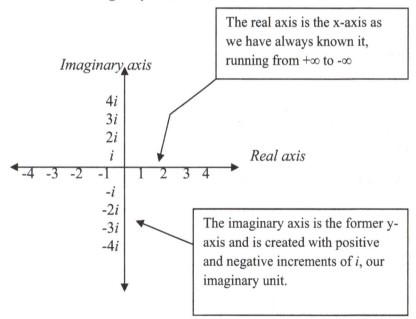

The real axis is the x-axis as we have always known it, running from $+\infty$ to $-\infty$

The imaginary axis is the former y-axis and is created with positive and negative increments of *i*, our imaginary unit.

Looking at the first quadrant, let's plot the complex number $4 + 3i$. We plot the real part (*a*) on the real axis, and the imaginary part (*bi*) on the imaginary axis

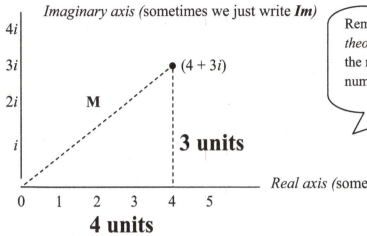

The magnitude of a number is defined as its straight-line distance from the origin. To find the length of **M** (above), we use the Pythagorean Theorem.

Since $M^2 = 4^2 + 3^2$ then $M = \sqrt{4^2 + 3^2} = \sqrt{25} = 5$

The magnitude of 4 + 3i, written as follows… $|4 + 3i| = 5$

This tells us that **the magnitude of a complex number is a real number.** We can prove this in general using a graphical approach in the picture above. You're probably thinking, why do we do everything "in general." Mathematicians are funny that way. We would rather prove something once using representative letters that are good for every number in the domain, instead of doing an infinite number of calculations with numbers to prove the same thing…think about it!! {{As Napolean Dynamite says, "I made like affinity of those at camp!}}

From the graph above you can see that the horizontal leg of the triangle (regardless of the length) is always "a," and the vertical leg of the triangle (again, regardless of length) is always "b" imaginary units.

And so the magic words…..*In General*, we have…

$$|a + bi| = \sqrt{a^2 + b^2}$$

Since a and b are both real numbers, the square root of those squared is also a real number.

Ah Hah!! Now we know how the complex numbers were sneaking in to the real number system. They were *surreptitiously* disguising themselves as their own magnitude (a real number), and slipping by the all powerful security systems.

Warning – another big word alert

Professor's Practice Pause

Instructions: For each of the complex numbers below, plot the number on the complex plane, and find the magnitude.

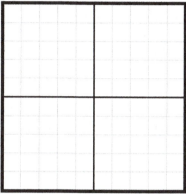

1. $5 + 12i$

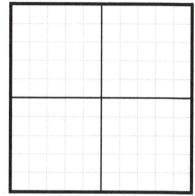

2. $-3 - 4i$

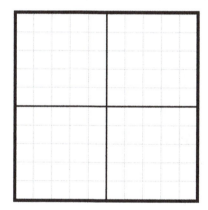

3. $0.56 + 1.7i$

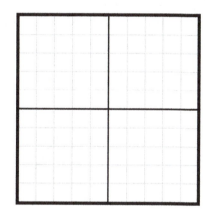

4. $(-1/2) + (3/4)i$

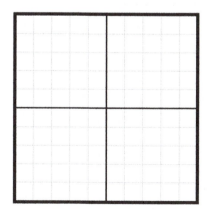

5. $0 + 6i$

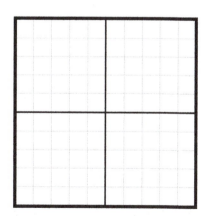

6. $10 - 0i$

Associated End of Booklet exercises are 1, 2, 4 and 5

Our Story Continues...The dreaded Doppelgänger ...

There are times in life when we think we have pulled something off – you know, "gotten by" with something we shouldn't have. And usually, the consequences of those actions take time to catch up with us. Such is true with our supposed antagonists (the sneaky complex numbers that snuck in). It was too late before they realized **2 very bad consequences** of their illicit activities.

One...They could never return to the complex plane from whence they came...and...

Two...They had each unknowingly brought with them their evil, heinous, fowl, putrid twin...the Doppelgänger ...(dun! dun! dahhhhhhhh!...evil music sounds in the background)

Before we become too afraid, we look at the first consequence...

Why could they not return? Going *from* a complex number *to* a magnitude is a ***function*** In other words, there is only one magnitude for every complex number, as you have seen in your Practice Pause. But, let's try and figure out how many complex numbers can be made given one magnitude.... *It's Geometry Time*!!

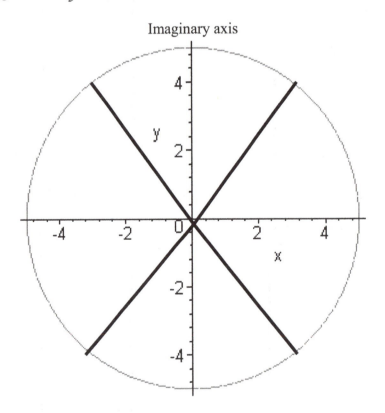

In the graph of the circle above, we have embedded the magnitudes of 4 complex numbers, all having magnitude of 5. In fact, any point on the outside of the circle represents a complex number. All of them have a radius (magnitude) of 5 units.

Our unsuspecting friend, $4 + 3i$, came over as "5", but it now has an infinite number of complex numbers that it can become on the home trip. The probability that it will find its original a and b values is basically 0…*it has lost its identity*! Doomed it is, forever to walk the streets of the Real number system, longing to have its old identity back. The Author has only three words for this…Crime Doesn't Pay!

Number two…

The theory of the Doppelgänger states that there is a person exactly like you somewhere else on the planet. If you were to meet up with that person somehow, you would both disappear – "like mahter n' unti-mahter, Captain." The numerical center of our story, $4 + 3i$, brought with it its Doppelgänger – its "opposite" if you will.

The Buddy system…

Do you remember 5th grade camp in school? You slept in those decrepit, old cabins, woke up too early, ate nasty food, went to bed too late, were hounded by the camp bully, and you had a counselor named Lars who was an ex-Marine. Forgive me for evoking those terrible memories. I've noticed in all such 5th grade camps, there is always a pond that the counselors tell you to stay away from. In fact, they usually say, "If you have to go down to the pond, use the Buddy system!" Complex numbers use the buddy system. A complex number and its 'buddy' (opposite, Doppelgänger , or whatever else we previously addressed them as) are actually called **complex conjugates**.

Here is a definition…

> The complex conjugate of $a + bi$ is $a - bi$.

Wherever there's an $a + bi$ you'll always find an $a - bi$. You'll never see just one complex number walking down the street – there will always be its complex conjugate as well.

> Examples of complex conjugates…
>
> $2 + 3i$ and $2 - 3i$
>
> $-2 - 0.75i$ and $-2 + 0.75i$
>
> $Q + Pi$ and $Q - Pi$

To form a complex conjugate given a complex number you simply *switch the sign* between the real and imaginary parts (a and bi). We will see how this comes into play when we are solving for complex roots of polynomial equations in the Booklet on Quadratic Functions.

Back To Our Story....

Here they were…all these complex numbers (and now all of their complex conjugates as well), stuck forever in the Real number system, without a chance of getting back home. And just like the beginning of our story, time passed on. More numbers mingled. Real numbers began going to the complex number system. And, finally, one day (a day much like today, in fact), all numbers were united again – large and small, positive and negative, rational and irrational, and (of course) real and imaginary. The novelty of the Worm Hole wore off. Instead of all these fantasy-type names, we began using Magnitude and conjugate. Things became….as they are today.

The End.

The Algebra of complex numbers

(If you get some Complex numbers for Christmas, what will you do with them?)

They may look strange, but complex numbers are just that – numbers. As a kid, you learned how to add, subtract, multiply, and divide numbers, right? [And if you didn't, you most certainly learned it again in Booklet on Real Numbers of this crazy text!]

Well complex numbers want some attention as well, so we should look at the arithmetic and algebra of the complex plane.

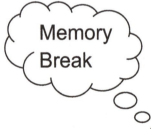 The phrase *combine like terms* sticks out in our heads, just like *cross-multiply* and *same – change – flip*, and many others. Combining like terms (CLT, for short….not BLT because that's breakfast) is also a method used for **adding and subtracting complex numbers**.

Imagine you are counting a pile of coins. You empty your jar of coins on the table and begin counting. Or, perhaps you are a pile person…first you make piles of pennies, piles of nickels, and so on, and then you count. Well, this is how we do arithmetic with complex numbers. We add the real parts together and add the imaginary parts together. Hence the phrase, combine like terms.

$$a + b\,i + (c + d\,i) = a + c + (b + d)\,i$$

$$a + b\,i - (c + d\,i) = a - c + (b - d)\,i$$

> **Adding** and **subtracting** complex numbers

Multiplying complex numbers is done by using our friend FOIL from the Booklet on Basics. Remember, FOIL stands for "Fools Oppose Intelligent Learning." No, that wasn't it…

Acronym Exercise: With the permission of the nearest adult, make up a definition for the acronym *FOIL* that somehow relates to education…place answer on line below

_____

"First, Outer, Inner Last," …that's what it was! We used the Augsburger Initials font, because we believe it may have been Martin Luther that quoted this first – although we may be mistaken. OK! We can do this, but we need to first investigate multiples of i.

We know that $i = \sqrt{-1}$

$$i^2 = \sqrt{-1}\sqrt{-1} = -1$$

$$i^3 = i^2 i = (-1)i = -i$$

$$i^4 = i^2 i^2 = (-1)(-1) = 1$$

> From the left, you can see that we have a repeating pattern of i, -1, -i, 1, and so on since i^5 is the same as i. The most important multiple to keep in mind is that $i^2 = -1$.

Let's use our two complex numbers from addition and multiply the together…

$$(a+bi)(c+di) = ac + adi + cbi + (bi)(di)$$

which, when simplified, gives…

$$ac + [ad + cb]i + bdi^2 \quad \text{…and since… we have} \quad ac + [ad + cb]i - bd$$

Or simply… $\underbrace{ac - bd} + \underbrace{[ad + cb]i}$

Real part   Imaginary part……….

When you multiply 2 complex numbers you (almost always) get a complex number.

Example of multiplication…

$(\mathbf{3} + 7i)(\mathbf{-2} - i) =$

FIRST = 3* (-2) = -6

$(3 + 7i)(-2 - i) =$

LAST = 7i*(- i)
    = -7i²
    = (-7)(-1) = 7

$(\mathbf{3} + 7i)(-2 - \mathbf{i}) =$

OUTER = 3*(-i) = -3i

Putting all 4 terms together, we have
(-6 +7) + (-3i – 14i) = **1 – 17i**

$(3 + \mathbf{7i})(\mathbf{-2} - i) =$

INNER = 7i*(-2) = -14i

Notice the answer is in complex form

Can you multiply a complex number and get a real number? Yes! Do you remember the story? What happens when a person meets its' Doppelgänger ? They cancel each other out. Let's look at an example of multiplying a complex number by its complex conjugate….and let's do it …(Oh no! Not that again)…yes – *In General* (Ahhhhhhhhhhhhhhhhhhh!!!)

We choose an unsuspecting complex conjugate pair…$(a + bi)$ and $(a – bi)$

$(\mathbf{a} + bi)(\mathbf{a} – bi) =$ $(a + \mathbf{b}i)(a – \mathbf{b}i) =$

    FIRST $= a* a = \underline{a^2}$ LAST $= bi*(-bi)$
                                              $= -b^2 i^2$
                                              $= (-b^2)(-1) = \underline{b^2}$

$(\mathbf{a} + bi)(a – \mathbf{b}i) =$

    OUTER $= a*(-bi) = \underline{-abi}$ Putting all 4 terms together, we have
                                       $(a^2 + b^2) + (-abi + abi) = \mathbf{a^2 + b^2}$

$(a + \mathbf{b}i)(\mathbf{a} – bi) =$

    INNER $= bi*a = \underline{abi}$ Notice the imaginary terms drop out
                                     The answer is a real number

---

Division of complex numbers is the longest operation, but it combines just about everything else that you have done so far…addition, subtraction, multiplication, and even conjugates. Let's call on our original pair of complex numbers…. *Vanna - the numbers please*……Tah – dah! $a + bi$ and $c + di$

Dividing them, we get…. $\dfrac{a + bi}{c + di}$ Well…..what do we do next??

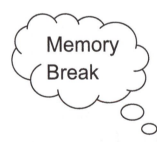

Memory Break

How do you get a *common denominator*? Multiply the top and Bottom by the same factor, right. We kind of do the same thing Here, but we use the complex conjugate of the denominator…. Like so --

$$\left[\frac{a+bi}{c+di}\right]\left[\frac{c-di}{c-di}\right]$$

And how do we *multiply fractions*? Multiply across the top and Multiply across the bottom.

$$\frac{[(a+bi)(c-di)]}{[(c+di)(c-di)]}$$

In the numerator, we FOIL, and we know the answer already to the denominator...$c^2 + d^2$

$(a + b\,i)(c - d\,i) \;=\; [a\,c + b\,d] + (b\,c - a\,d)\,i$ <u>numerator</u>

$(c + d\,i)(c - d\,i) = c^2 + d^2$ <u>denominator</u>

Putting them together gives us... $\dfrac{[a\,c + b\,d] + (b\,c - a\,d)\,i}{c^2 + d^2}$

Separating fractions gives us this answer.. {{*remember*...(A + B) / C = A / C + B / C}}

$$\frac{a\,c + b\,d}{c^2 + d^2} \;+\; \frac{(b\,c - a\,d)\,i}{c^2 + d^2}$$

And since a, b, c, and d are all real numbers, we have "a + bi" form.
Another way to say it is (number) + (number)i.

Take a few minutes right where you are to practice. This is kind of like remembering something that was eluding your memory – like a dream or a poem. As soon as you remember it you should write it down. As soon as you learn a new math concept...try it out....like a rehearsal.

Professor's Practice Pause

Multiply and divide the following two complex numbers...$7 - 6i$ and $-2 + 3i$

A) $(7 - 6i)(-2 + 3i)$

B) $\dfrac{7 - 6i}{-2 + 3i}$

Associated End of Booklet exercises are 3, 6 – 11

Although we tend to be mathematical purists, and think that mathematics has a gracefulness all its own, which doesn't need to be tainted with applications, we also realize that it helps students to understand where some of these topics play out in life. Even though complex numbers are used extensively in follow-on math courses, we should also ask the question, **"Where can *we* find some complex numbers?"**

Application #1: Electrical Circuit Analysis has used these numbers quite a bit. Even fairly simple RLC circuits (those containing a combination of resistors, inductors, and capacitors) have imaginary values of *Impedance* and *Frequency Response*. Some of the techniques used to solve for the voltages and currents use complex numbers (this is also called Phasor notation)....Seriously ... no Star Trek joke!

QuickSearch: Use Google, Wikipedia, or whatever search engine you employ and look up RLC circuits, Phasor notation, and Impedance … see what you find.

Application #2: Almost the same concept is true in the study of motors in electromechanical energy conversion. Most motors (because they have some of those same electrical and magnetic characteristics like circuits) have complex components of torque.

QuickSearch: Again use your favorite search engine to look up induction motor, complex torque, or squirrel-cage induction motor

Application #3: Complex numbers can be used to generate Fractals. Using complex numbers as *seeds* in Julia and Mandelbrot sets generate different fractals.

QuickSearch: Search for Fractals, Mandelbrot set, Julia set, Fractal mathematics, Fractals in nature, etc…

Application #4: (Most applicable to us) Complex numbers will show up as solutions to some polynomial functions, as we will see in the Booklets on the Quadratic family of functions and Higher Order polynomials.

Complex Number operations on the TI-83

The TI-83 (and newer models) have the ability to do manipulations with complex numbers. We have to start by putting the calculator in *complex mode*.

To do this, press the **MODE** button and scroll down to Real and then over to a+bi. Press **ENTER** and you should see a+bi selected. Now your TI is in complex mode. You can leave it in this mode for the entire course…real-valued calculations are not affected.

Addition & Subtraction

On the home screen, type in (5+4i)-(3-2i) and press **ENTER**. The *i* button is **2nd** then **Period (·)**. Your TI should display 2+6i.

Multiplication

Press **2nd** then **Enter** and you will see, again, (5+4i)-(3-2i), however this time it is in Edit Mode. Use the arrow key to change + to *. Press **Enter** and your TI displays 23+2i.

Division

Press **2nd** then **Enter** once more to see (5+4i)*(3-2i). Use the arrow key to change * to ÷. Press **Enter** and your TI displays .538+1.692i.

Magnitude

To find the Magnitude of a complex number, press the **MATH** key and arrow to **NUM**. Press 1:abs. Insert 3+4i and close the parentheses. Press **ENTER** and the TI displays 5...we knew that.

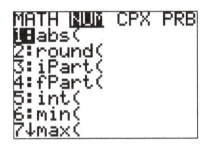
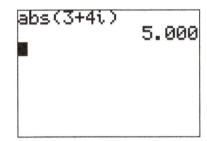

Converting negative square roots to complex form

Convert √-85, √-100, and √-4.6 to complex form (a+bi). On the TI, press **2nd**, x^2, **(-)**, **85**, **)**. Press **ENTER** and the TI displays 9.220i. Do the same for √-100, and √-4.6.

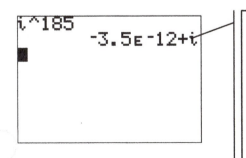

TI-83 NOTE...Type i^185 and press **ENTER**. Your answer may seem strange...

What's this -3.5E-12+i all about?

-3.5E-12 is in scientific notation. It really means -3.5 x 10^{-12}. This means we move the decimal place to the left by 12 places...-0.0000000000035. This is your TI-83's way of saying 0. The answer is simply 0 + i or just i.

i^185 = i.

Topical Summary of Complex Numbers

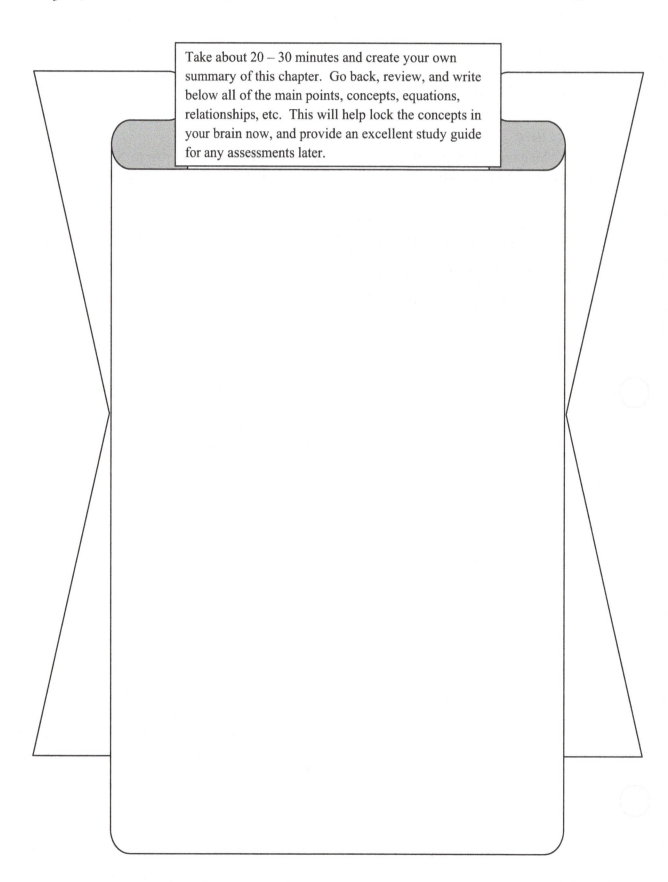

Take about 20 – 30 minutes and create your own summary of this chapter. Go back, review, and write below all of the main points, concepts, equations, relationships, etc. This will help lock the concepts in your brain now, and provide an excellent study guide for any assessments later.

End of Booklet Exercises (you knew they were coming)

1) Define the imaginary unit i.

2) Re-write the following numbers in simplest <u>radical</u> form using i. ((Remember the *factor tree* from the Booklet on Basics))

(a) $\sqrt{-1}$

(b) $\sqrt{-2}$

(c) $\sqrt{-4}$

(d) $\sqrt{-81}$

(e) $\sqrt{-45}$

(f) $\sqrt{-72}$

(g) $-\sqrt{-20}$

(h) $3 + \sqrt{-20}$

$(\sqrt{-1})$ $6 - \sqrt{25}$

(j) $\sqrt{-x^2}$

(k) i^{196}

3) Using your TI-83, calculate i^{225} _____.

(a) Explain how you could find this answer relatively quickly without the calculator.

(b) Use your method to find i^{812} , and check your answer with the TI-83. Do the two answers agree?

4) Graph the equation $y = x + 3i$ in the complex plane, where x can be any real number. What does your graph resemble?

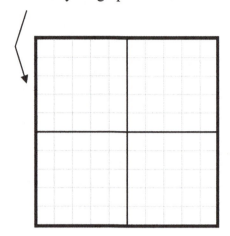 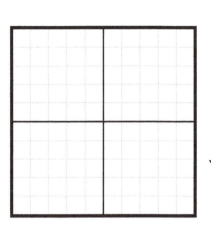

5) Find the complex number in the first quadrant with a *magnitude* $\sqrt{20}$ and real part of a = 2. Find the same number in the fourth quadrant.

6) Simplify the following expression to $a + bi$ form…$-3 - 5i + \sqrt{-16} + 3i - 12$

7) Find the complex conjugate of the number given and then multiply the number by its conjugate.

      (a) $6 - 4i$ conjugate = _____

      (b) $-2 + 9i$ conjugate = _____

(c) What do you notice about both of your answers to the multiplication parts of (a) and (b)? What trend does this suggest in general? (Yes, we said, "In general!!")

8) Three different complex numbers take the Worm Hole and end up having the <u>same</u> value. Write a possible group of three complex numbers that can do this.

9) Perform the indicated operations…((Check your answers using the TI-83)

(a) $(0 - 5i) + (5 - 0i)$

(b) $(0 - 5i)(5 - 0i)$

(c) $(3 - 7i) - (-4 + 12i)$

(d) $\dfrac{-3 + i}{-3 - i}$

(e) $(1.3 - 7.45i) - (-4.9 + 12.02i)$

(f) $(2 - 6i)(1 - 8i)$

(g) $(3 - 2i) / (5 + 7i)$

(h) $(1 + i)^2$

$(\sqrt{-1})$ $\dfrac{(6 - i)^2}{6 + i}$

(j) $\dfrac{[(13 - 7.5\ i)(-6 + 12\ i)(i - i)]}{[(-i)(5 + 4\ i)]}$

10) In the Booklet on Quadratic Functions, we will learn how to solve quadratic equations that have complex answers using the quadratic formula. Part of the quadratic formula is called the *discriminant*. It looks like this…

$$\sqrt{b^2 - 4\,a\,c}$$

Write the discriminant using the imaginary unit i if a = 2, b = 3, and c = 4

11) Find the magnitude of the following complex numbers. Check your answers with the TI-83.

(a) $-1 - 5i$

(b) $10 + 3i$

(c) $2.7 + 8.4i$

(d) $x + 2i$ [you cannot check this one]

Thinking & Research Questions

1) Conduct some research (internet or otherwise) on the history of the imaginary unit, i. Summarize your research in a paragraph.

2) Based on your research in question 1), how do you think the term "imaginary" has troubled algebra students throughout history? If you could rename the imaginary numbers, what would you call them, and why?

3) The reactive capacitance (complex impedance) of an RC circuit (that is, a circuit with resistors and capacitors) is given by $X_c = -\dfrac{i}{\omega C} = \dfrac{1}{i \omega C}$ ... where ω is the frequency in radians and C is the capacitance in farads. (Note: the symbol ω is the Greek, lowercase letter omega).

   (a) Show that $-i$ (in the middle part of the equation) is equal to $1/i$ (from the last part). (Hint: if $i^2 = -1$, then $-i^2 = 1$)

   (b) If $\omega = 2\pi$ and C = 0.1, find X_c in a + bi form.

Booklet on Equations and Expressions

The vocabulary of equations and expressions

Math is a language! I think you've heard that somewhere before, no?

Some Definitions...

An equation has only three things…an equal sign (=), something to the left of the =, and something to the right of the =. So really, all equations look like this:

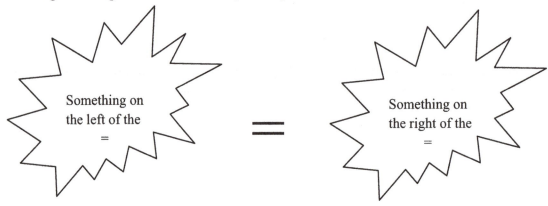

Something on the left of the =

=

Something on the right of the =

The equation can be as simple as $1 + 1 = 2$, or as difficult as $5x^2y^3 - 15xy + 1 = -10.45x^2y^2 - 7$.

Some equations can be solved – we will do this as we progress!
Some equations cannot be solved – even by the smartest mathematicians around!
Some equations represent fundamental truths, called *identities* … $2 + 2 = 4$ is an example.

So when we talk about <u>solving an equation</u>, we are trying to find the one (or sometimes more than one) number that we can substitute for the given letter to make the equation **true**, or **balanced**. If you want to think of the equation like a scale that must have the same weight on both sides in order to be balanced.

An equation is like a balance beam or scale…

Math to English
Notice that *solving* is the <u>verb</u> that goes with the <u>noun</u> *equation*.

An equation is solved, balanced, or true, when the left side is equal to the right side, like a scale. Perhaps you have heard the phrase *Lady Liberty's scales are tipped in favor of…*talking about the *inequality* of some judicial matter. In this case the scales of justice are not balanced, not true, not equal. The same can be considered for a math equation.

A solution is reached when the equation is in this form: variable = number…like x = 3.

Example: Is x = 5 the solution to the equation $3x - 15 = 0$?
In other words, if we substitute 5 for x, will the left side equal the right side?

$$3(5) - 15 = 0$$
$$15 - 15 = 0$$
$$0 = 0…………Yes!\text{ So } x = 5 \text{ is the solution.}$$

Example: Is r = 2.68 the solution to the equation $-6r + 12 = -4.02$?

$$-6(2.68) + 12 = -4.02$$
$$-16.08 + 12 = -4.02$$
$$-4.08 = -4.02……………No!\text{ They are close, but not equal. So, } r = 2.68$$

is *not* a solution to that equation.

What are the two things on both sides of the equal sign? Do they have names? Why yes, as a matter of fact they do! They are called *expressions*. Hey that brings back memories of that great Motown song, *Express yourself, express yourself*!!……woops…..sorry.

Like equations, expressions can be very simple (1 + 1, x + 2, y, etc.) or very difficult as in the example used above ($5x^2y^3 - 15xy + 1$ and $-10.45x^2y^2 - 7$). We don't solve expressions, we *simplify* them. Other verbs can be used as well, such as reduce or substitute, but typically we are simplifying.

Math to English
Notice that *simplify* is the <u>verb</u> that goes with the <u>noun</u> *expression*.

Terms are used to identify the parts of an expression. A term is nothing more than a bunch of letters and numbers stuck together using multiplication. Terms are separated by + and – signs. Here's an example of a term and its components…

$$8x^2y^3$$

A term's *coefficient* is the number before the letters. 8 is the coefficient of the term above.

The *variables* are the letters…x and y are the variables in that term. Another way to think of a variable is to consider it just a temporary storage location (like a shoebox) that you are eventually going to fill with a specific real number (or pair of shoes).

Variables have *exponents*. As we learned before, the exponents are superscripts (super meaning *above*, and script meaning *writing*, so a superscript is *writing above* something). The exponents above are 2 and 3 (x squared and y cubed).

An expression could also have a _constant term_ (or simply constant). A constant is a number out on its own; it never changes. 3 can never become -7, no matter how hard it tries. What are the constant terms in the equation shown at the beginning of this Booklet? (also below)

$$5x^2y^3 - 15xy + 1 = -10.45x^2y^2 - 7$$

If you said 1 and -7 you are correct! The 5, -15, and -10.45 are the coefficients. There are 3 terms in the left-hand expression and 2 terms in the right-hand expression.

Example: In the given equation, list the terms, variables, coefficients, constants, and exponents. $\qquad p^2z^3 - 12zy + \pi = 23p^4y^2 - 1$

The terms are p^2z^3, $-12zy$, π, $23p^2y^2$, and -1 of which π and -1 are constants.

The variables are p, z, and y. Notice that p^4 and p^2 are the same variable even though they have different exponents.

The different exponents are 1, 2, 3, and 4.
Remember that the exponents of z and y (in the term $-12zy$) are actually 1. We could rewrite the term as $-12z^1y^1$, but we don't usually do that in math…it is understood.

It's time for the *Professor's Practice Pause*

Remember the tree diagram you constructed about the <u>hierarchy</u> of real numbers? (see the Booklet on Real Numbers)

Warning – big word alert

Well, I would like you to create another tree diagram on the hierarchy of *the Equation*. In groups of 2 or 3 take the words and symbols below and organize them into a tree diagram representing the different parts of an equation.

<u>Words / Symbols for your Tree Diagram</u>:

Expression, =, Terms, Equation, constants, variables, exponents, + and – signs, coefficients.

Here is a hint: $-5x^2y + 2x - 7 = 3y^3 - 2x + e$

Associated End of Booklet exercises are 1 – 8, 13 – 20

Solving simple equations (in one variable)

If I ask you to get $15 worth of gasoline, and the price at the station is $1.50 /gallon (*Oh yes, I am dreaming of way better times. Actually, I can remember when gas was fifty cents per gallon…with that information you could probably find my age!*). Anyway, you "automatically" know that you are going to be able to get 10 gallons. Without even realizing it, you solved an equation for the number of gallons (variable), given the amount of money you have and the price per gallon (constants). At the publication of this textbook, you will probably need about $35 for those same 10 gallons ☹

This specific equation looks like this: $15 = 1.5 x$, where x is the number of gallons.

That's an easy one…but what if the equation looked like this: $1.73x - 15.9 = 245.7$. Now it isn't so obvious what x has to be to make that equation true is it? This is why we need a system of steps or principles to systematically find that specific x value. You have probably been taught millions of ways to do this, but I am going to boil those steps down to two basic principles, the "Golden Principle" and the "Silver Principle." That's it – just **two basic principles** upon which hinge the solving of any equation.

The Golden and Silver Principles

The Golden Rule says "Do unto others as you would have them do unto you." That's not exactly what we are doing here, although it would be good to follow that rule.

The <u>Golden Principle</u> says, **"<u>Whatever</u> you do to one side of an equation, you <u>must</u> do to the other side."**

As a math student, you are free to do whatever you want to one side of an equation. For example, you might add something to it, subtract from it, multiply it by a number, divide it by a number, raise it to an exponent (power), flip it over, paint it red, whatever. However, as soon as you decide to do something to one side, you *MUST* do the same thing to the other side.

Think of an equation as that balanced set of scales that we talked about. You know that if you add weight to one side, you will have to add weight to the other side to make them balanced again.

<u>We do algebra to whole sides…not terms!! I can't emphasize this enough! Every term in an expression must receive the same, algebraic treatment.</u>

Example: Let's start with the equation $y = 3x$. The following table shows two things: a choice of a mathematical operation applied to both sides, and the resulting <u>equivalent</u> equation. In other words, the resulting equation describes the same relationship between the two variables, just in a slightly different form.

If you need to reassure yourself of this, use the values $y = 3$ and $x = 1$, and verify that they make every equivalent equation true, as well as the original.

ENGLISH	**MATH**
Operation	Equivalent Equation
Add three to both sides	$y + 3 = 3x + 3$
Subtract 10 from both sides	$y - 10 = 3x - 10$
Divide both sides by 25	$y / 25 = (3x) / 25$
Multiply both sides by a variable, z	$yz = (3x)z$
Square both sides	$y^2 = (3x)^2$
Flip both sides over	$1/y = 1/(3x)$

Yet another synonym for solve:

> ### Isolate
>
> *Solving* for x also means isolating x. That is, getting x all by itself on one side of the equation and getting the numbers on the other.

The <u>Silver Principle</u> says, **"In order to <u>isolate</u> a variable, do the <u>opposite</u> operation."**

Why the opposite? To undo addition, we apply subtraction. To undo multiplication, we apply division, etc. In math, we call these *inverse* operations.

Example 1: In the equation $x + 3 = 8$, since the 3 is *added* to the x, the silver principle tells us to *subtract* 3. But if we subtract three from one side, we have to subtract it from the other side also (golden principle).

$$x + 3 = 8$$
$$\underline{- 3 \quad -3}$$
$$x = 5$$

Example 2: Solve for p in the equation $5p - 25 = 10$. We want to isolate all the "baggage" away from the variable p. Since the 25 is being subtracted, we would add 25 (to both sides). The p is then being multiplied by 5, so we would divide (both sides – meaning every term in that expression) by 5. The sequence looks like this:

Step 1 $5p - 25 = 10$
$$\underline{+ 25 \quad + 25}$$ (note that $-25 + 25 = 0$)

Step 2 $\dfrac{5p}{5} = \dfrac{35}{5}$ (note that 5 divided by 5 is 1, 1 times p is p)

Step 3 $p = \dfrac{35}{5} = 7$

Example 3: Solve for y in the following equation, $5/y + 7 = 17$. Notice the variable (y) is in the denominator. The sequence would look like this:

$$5/y + 7 = 17$$
$$\underline{\quad -7 \quad -7 \quad}$$ (subtract 7 from both sides)

$$5/y = 10$$ (next, flip both sides over to get the variable on top)

$$y/5 = 1/10$$ (next, multiply by 5 to "undo" the division)

$$5(y/5) = 5(1/10)$$ (5 divided by 5 is 1, 1 times y is y)

$$y = 5/10 = 1/2 = 0.5$$

Example 4: Solve for the variable t in the following equation, $5(t + 3) = 60$. Notice that the t is inside a set of parentheses. The order of operations (from the "Basics" Booklet) states that we have to take care of parentheses first. So in this problem we will have to apply the *Distributive Law*. The distributive law tells us to multiply *everything* inside the parentheses by whatever is on the outside (5, in this example). The sequence would look like this:

$$5(t + 3) = 60$$ (use the distributive law)

$$5t + 5(3) = 60$$

$$5t + 15 = 60$$ (subtract the fifteen from both sides)
$$\underline{\quad -15 \quad -15 \quad}$$

$$5t = 45$$ (divide both sides by 5)

$$\frac{5t}{5} = \frac{45}{5}$$ So, t = 45 / 5 = 9

Check: $5(9 + 3) = 60$
$5(12) = 60$
$60 = 60$ check!

Check, Please!...√...√...√...

Math is unlike any other topic because you can check almost every problem you ever work on. Your feedback in math can be immediate. You can't do that in English, right? You can't just write a paper and say, "Oh, I think I'm going to get this part correct, but this part wrong..." In math you can know right away by checking your answers. In Example 3, if we replace the y in the original equation by our answer (0.5), the equation should be balanced. 5 divided by 0.5 is 10 and 10 plus 7 is 17. It checks out!! You should always get in the habit of checking answers when you are done – it's good math hygiene...just like flossing is good dental hygiene, and reading is good mental hygiene.

Do math neatly!

Time for me to rant and rave a little….

Have you ever turned in an English paper like the one below?

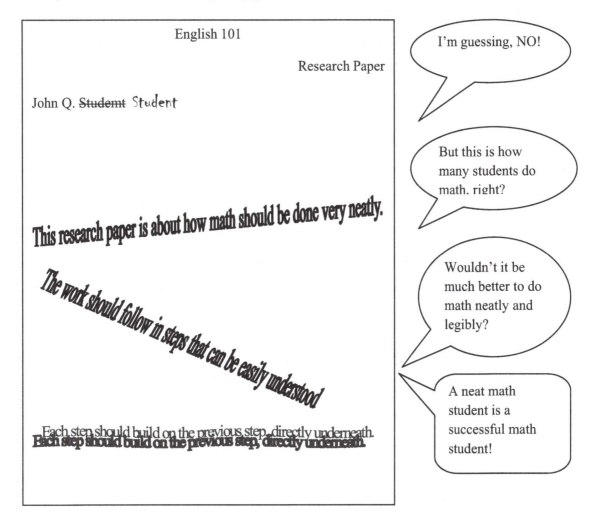

English 101

Research Paper

John Q. ~~Studemt~~ Student

This research paper is about how math should be done very neatly.

The work should follow in steps that can be easily understood

Each step should build on the previous step, directly underneath.
~~Each step should build on the previous step, directly underneath.~~

I'm guessing, NO!

But this is how many students do math. right?

Wouldn't it be much better to do math neatly and legibly?

A neat math student is a successful math student!

Do your math step-by-step, one line under the other line, continuing your mathematical thoughts just as if you were continuing your English thoughts in a research paper. *"…line upon line, precept upon precept."*

Notice how I solve the equation step by step, only changing one thing per line, and re-writing everything else that did not change.

$$5/y + 7 = 17$$
$$\underline{\quad -7 \quad -7\quad}$$ (subtract 7 from both sides)
$$5/y = 10$$ (next, flip both sides over to get the variable on top)
$$y/5 = 1/10$$ (next, multiply by 5 to "undo" the division)
$$5(y/5) = 5(1/10)$$ (5 divided by 5 is 1, 1 times y is y)
$$y = 5/10 = 1/2 = 0.5$$

Notice the commenting above…A very good idea, and something you should do as well.

Klingons and Danglers

By now you must have figured out that I'm a trekkie – a person who has watched, like, every Star Trek® episode and movie. I don't go to conventions or stuff like that – that's a little *too* weird, you know what I mean? So, I like to use Star Trek words and examples. Besides, everyone knows what a Klingon is anyways! A Dangler was one of those snake-like, hanging creatures in episode 163. You'll be amazed at how I apply these to mathematics…

Let's go back to the equation in example 2: $5p - 25 = 10$

We first got rid of the 25 by adding it to both sides. Could we have first divided by 5? Yes, we could have, and that would have been algebraically correct, but watch what happens when we do:

Step 1 $\dfrac{5p - 25}{5} = \dfrac{10}{5}$ (Divide both **<u>sides</u>** by 5)

Notice, algebra is done to **whole sides**. An entire side is like the scales on the balance beam We now have a compound fraction, one that looks like $\dfrac{A - B}{C}$.

$$\boxed{\dfrac{A - B}{C} = \dfrac{A}{C} - \dfrac{B}{C}}$$

Step 2 $\dfrac{5p}{5} - \dfrac{25}{5} = \dfrac{10}{5}$ (Break up the compound fraction)

Step 3 $p - 5 = 2$ (simplify each term)
                        $\underline{+ 5\ +5}$ (add 5 to both sides)

Step 4 $p = 7$

We get the same answer, but we have to do a bit more work. As the complexity of the equations increase, the complexity of solving them this way also increases. That's why I recommend taking care of the Danglers before the Klingons….what??

Consider the following *psycho-social* analogy:

> We all have people in our lives that aren't too close. They're just sort of *dangling* on the fringes of our lives. These people are very easy to dispense with… a phone call, quick email, and it's done. However, there are others who want to suck the very life-blood from our existence – "*klinging*" tightly to our every move. They are more difficult to get rid of. Hence…the Klingons and the Danglers.

In our example above, 5p – 25 = 10, the –25 is the Dangler – a little further away from the variable, while the 5 is the Klingon – stuck tight to the variable p.

The moral of this story is to get rid of the Danglers before the Klingons.

Example: When solving the equation $Kx + D = $ Answer, it is typically easier to move the D away from the x <u>before</u> you move the K away from the x.

TI PRACTICE TIME

The TI-83 can also solve equations, believe it or not, but it doesn't solve them algebraically, it solves them graphically. Remember, the calculator is dumb, but *fast*. It has been programmed with different types of *algorithms* that allow it to look like it is doing symbolic algebra – but it isn't.

> Warning – big word alert

In fact, if your professor allows you, I have one of those algorithms in the Booklet on Quadratic functions that you might find useful. I wrote a TI-83 program to solve for the roots of any quadratic equation. If that sentence didn't make any sense to you, that's ok – it will when we get there. In any case, you are free to use that program if you are allowed.

An important concept to understand… - the relationship between an equation and its graph. The relationship is that they are the same. From the equation, we can determine the algebraic representation of the solution, and from the graph we can determine the graphical representation of the solution. Let's consider the equation *3x + 4 = 7*.

<u>Algebraic representation</u>

$$3x + 4 = 7$$

The solution occurs at the value of x = 1. This value makes the left side *equal* to the right side.

Important concept

<u>Graphical representation</u>

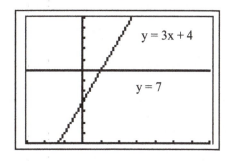

$y = 3x + 4$

$y = 7$

The left side graph (y = 3x + 4) *intersects* the right side graph (y = 7) at the value of x = 1.

<u>The x-value at which the expression on the right equals the expression on the left is exactly the same point that the graph on the right intersects the graph on the left.</u> An equation is an algebraic representation of the solution and graphs are a graphical representation of the same solution.

So how do we actually do this on the calculator? I'll show you.

Example: Let's solve the equation -5x − 12 = 27 using the TI-83.

Press **Y=** and enter the left expression (-5x − 12) into Y_1. Enter the right expression (27) into Y_2. Your screen should like this..

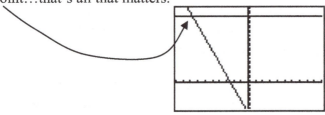

In order to assign a value as negative you must use the negative key (–). To subtract, you must use the subtract key  The 5x term is negative, but 12 is being subtracted.

Now adjust the window. We know that we will at least have to see a height of 27, so a YMAX value of 30 should be good. Experiment with the window values. The important thing to remember is that you must SEE the intersection before the TI can find it.

After you have a good window, press **Graph**. My window is good because I see the intersection point…that's all that matters.

Now, we use what I call the *6 magic buttons*…**2ⁿᵈ**, **TRACE**, **5:**, **ENTER**, **ENTER**, **ENTER**. Notice that the word after the **5:** button is intersect. This is also called the Intersect Method. Notice that before the first Enter, your TI shows First Curve? By pressing **ENTER** you are saying that it is correct. The same with Second Curve? Then it says Guess? Well we don't want to guess, we just want the answer, so we press **ENTER** the 3ʳᵈ time, and *viola*…

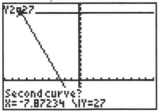

Our solution is x = -7.87234. The point of intersection is the point (-7.87234, 27).

Intersect Method (also called the 6 magic buttons)

1. Place the left expression in Y_1, and the right expression in Y_2

2. Find an appropriate window. The important thing is to visually see the point of intersection in your window.

3. Press **2ⁿᵈ**, **TRACE**, **5:**, **ENTER**, **ENTER**, **ENTER**

Professor's Practice Pause

Solve the given equation *algebraically* and verify your answer *graphically*. $8 - 2.5x = 17$

Algebraic work here…	Copy your TI graph here…

Associated End of Booklet exercises are 9 – 12, 21 – 37

Combining Like Terms (CLT)

Not BLT, that's bacon, lettuce, and tomato, and I could really go for one of those right now. Well, not *your* right now. When you read this, my *right now* will have come and gone!

Acronym Exercise: With the permission of yourself, make up a definition for the acronym *CLT* that somehow relates to your college…place answer on line below

_____

Solving an equation with many terms uses the same two principles, but it requires a little combining of like terms first. "Like Terms" are piles. There are two kinds of people in the world -- those who count loose change randomly, and those who count loose change by first organizing the coins into piles. CLT is the latter, you have a variable pile and a number pile.

Example: Solve the equation $10 - 1.5x + 26 - 7x = 14x - 9$.

Before we begin with the golden and silver principles we can <u>simplify the expression</u> on the left by CLT. I can regroup my terms using the following principle

$$A - B + C - D = A + C - B - D$$

I can combine terms in any order I want, as long as I maintain the correct signs for each term.

Re-writing gives... $10 + 26 - 1.5x - 7x = 14x - 9$

Now CLT... $36 - 8.5x = 14x - 9$ ...which side do you want the x's on? It doesn't matter, just choose one. Let's put the x's on the right, so that means we put the numbers on the left.

$$36 - 8.5x = 14x - 9$$
$$\underline{+\ 8.5x \quad\ +8.5x} \qquad \text{(add 8.5x to both sides)}$$

$$36 = 22.5x - 9$$
$$\underline{+9 \qquad\qquad +9} \qquad \text{(add 9 to both sides)}$$

$$45 = 22.5x \qquad \text{(now we divide both sides by 22.5)}$$

$$\frac{45}{22.5} = \frac{22.5x}{22.5}$$

$x = 45/22.5$ which we can easily see is **2**. If you can't easily see that, try the long division....

$$225 \overline{)\ 450}$$

Literal Equations

Literal equations have all or mostly *literals* (that is, letters or variables). Sometimes we call these equation *formulas*. The techniques for solving these equations are exactly, 100% the same as solving all the other ones we looked at. There is, however, one slight difference. We can combine numbers together, but we cannot combine letters together...A – B is A – B, there is no way around that. Here's an example of solving a literal equation...

Example: The area of a triangle is given by $A = \frac{1}{2} bh$. Solve the equation for h.

Currently the equation is solved for A, "A =." We want it solved for h, so we isolate the h.

$$A = \frac{1}{2} bh$$

$$2 \cdot A = 2(\frac{1}{2} bh) \quad \text{...multiply both sides by 2 to clear the fraction}$$

$$\frac{2A}{b} = \frac{bh}{b} \qquad \text{...divide both sides by b}$$

$$h = \frac{2A}{b}$$

Example: The perimeter of a rectangle with length L and width W is given by $P = 2W + 2L$. Solve this equation for L

$$P = 2W + 2L$$
$$\underline{-2W \quad -2W} \qquad \text{...subtract 2W from both sides}$$

$$P - 2W = 2L$$

$$\frac{P - 2W}{2} = \frac{2L}{2} \qquad \text{...divide both sides by 2}$$

$$L = \frac{P - 2W}{2} \quad \text{or} \quad \frac{P}{2} - W$$

We can see this graphically. If we cut the perimeter in half and then take away the width, we are left with the length...

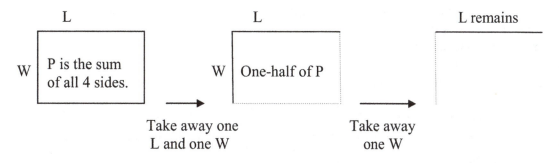

Your professor might just have this question on your next quiz...Shhhh! Don't tell!

A *literal* equation is?

    (a) One that should be used literally (i.e. word for word)

    (b) An equation with all or mostly variables

    (c) One that is easily read

    (d) Always representing a linear equation

Problem Solving

Everyone hates word problems for some reason?? I don't know why? *Life* is one big word problem, broken down into thousands of itsy-bitsy parts.

A very helpful man in this area is George Pólya. Aside from working in the fields of advanced Abstract Algebra and Geometry, George Pólya is considered to be the father of the problem solving technique (see his history at the end of this Booklet). For word problems, I like to break down his four steps even further: I present to you the *Modified Pólya Approach*.

(a) Read, read, and re-read the problem.
(b) Draw a picture, graph, chart, or other visual and label it with all information given.
    A visual in math is worth a thousand symbols ☺
(c) Figure out what you are being asked to find. This is the <u>unknown</u>. Assign the unknown a variable (letter). There may be more than one.
(d) Try to somehow relate the unknown to the given information using an equation or formula. Sometimes this requires an additional equation or relationship (area, volume, perimeter, distance, etc.)
(e) Solve the equation.
(f) Check your answer (if appropriate)

Example: As a **landscape specialist**, you are asked to provide a material estimate for some sod grass planting in a local corporation's flagpole area. The flagpole sits in the middle of the smaller of two <u>concentric</u> circles, and they would like to have the area in between the circles filled-in with sod grass. The inner circle (where the flagpole sits) has a radius of 6'. The outer circle has a radius of 18'. Sod grass costs you $1.75 / square foot. What price do you give them for the cost of material?

| What does concentric mean? |

Step 1: Read until you are familiar with what is being asked. You may have to do this several times.

Step 2: draw some sort of visual and label the picture

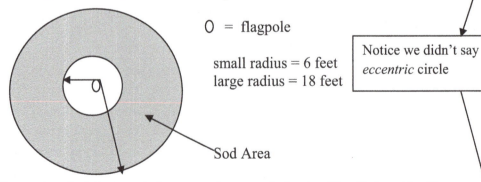

O = flagpole

small radius = 6 feet
large radius = 18 feet

| Notice we didn't say *eccentric* circle |

Sod Area

Step 3: We need to find the area in between the two circles (we'll call that A). Then we multiply by the cost per square foot of sod to obtain a price (we'll call that P).

Follow arrow next page

Step 4: The area of a circle is given by $A = \pi r^2$. To find A, we will subtract the small circle area from the large circle area so we just have the **green** sod area. To find P, we will multiply our result for A by $1.75.

> An eccentric circle has a lot of money, wears funny clothes and mostly hangs out with the Platonic solids.

A (outer circle) = $\pi (18)^2 = 3.14 \cdot 18^2 \sim 1017$ sq. ft
A (inner circle) = $\pi (6)^2 = 3.14 \cdot 6^2 \sim 113$ sq. ft

A (middle part) = A (outer circle) - A (inner circle) = $1017 - 113 = 904$ sq. ft.

Step 5: P = 904(1.75) = **$1582**…It would cost you $1582 to purchase the necessary sod.

In conclusion, you have to realize that every word problem or every problem where you are applying the problem solving technique will be slightly different. The content will be different; different visual representations will be drawn; the relating equations will probably vary from problem to problem. While these differences will exist, the general problem solving techniques do not. The best way to become proficient at solving problems is to…you guessed it…solve more and more problems.

Draw your own eccentric circle here…

Example: In Kyrgyzstan, they **rent camels**, not cars. The currency in Kyrgyzstan is the *Som*. The current exchange rate for US Dollars is $1 = 34.90 Kyrgyzstani Som (KGS).

Your mission, should you decide to accept it, is to rent a camel in Kara-Say and travel to Tash Kumyr to await further orders. At Isakovich's camel rental in Kara-Say you must pay 55 KGS per day for the camel **and** (believe it or not) 3 KGS per kilometer traveled. Use the map on the right to estimate the distance between Kara-Say and Tash Kumyr. You are told that the average Kyrgyzstani camel will travel 45 kilometers per day through the desert.

How much will the camel rental cost you in KGS?

How much will the camel rental cost you in USD?

(a) Estimate the number of kilometers between Kara-Say and Tash Kumyr and the number of days of travel this will require.

500 kilometers is a good estimate from the map, especially since you may not be traveling in a straight line between the two towns. 500 k / 45 k per day = 11.11 days. Since this is slightly more than 11 we should count on 12 days.

(b) How much will the camel cost us in KGS?

500 k · 3 KGS per k = 1500 KGS. 12 days · 55 KGS per day = 660 KGS. The total cost of the camel rental in KGS is 1500 + 660, or 2160 KGS

(c) How much will the rental cost us in US Dollars?

$1 = 34.90 KGS, so 2160 KGS = 2160/34.90, or $61.89. Now that's a bargain!! Everyone's happy! You're happy. Isakovich is happy. The camel is...*well*...maybe not.

Dimensional analysis

We used a concept called *dimensional analysis* above to work with units. We analyzed the dimensions (units). For example, in part (a) we calculated the number of days it would take us to travel by calculating the quotient of the number of kilometers and kilometers per day. Looking just at the units we have...

$$\frac{\text{kilometers}}{\text{kilometers per day}} = \frac{\text{kilometers}}{\frac{\text{kilometers}}{\text{day}}} \quad \text{or} \quad \frac{k}{\frac{k}{d}} = \frac{k}{1} \cdot \frac{d}{k} = d, \text{ the number of days}$$

Topical Summary of Equations and Expressions

Take about 20 – 30 minutes and create your own summary of this chapter. Go back, review, and write below all of the main points, concepts, equations, relationships, etc. This will help lock the concepts in your brain now, and provide an excellent study guide for any assessments later.

End of Booklet Exercises (Yes!...Once again)

1) Given the following equation, list out all of the "parts" on the lines below

$$-5x^2y^3 - 15z + 7 = \pi + z^3 - 12xz$$

Variables _____   Constant terms _____
Coefficients are _____   Non-constant terms _____

How many expressions are there?_____

2) Create an equation that has 4 terms, 2 different variables, with a coefficient of -6 and an exponent of 3.

3) Create an expression with 3 terms (one of them a constant term) and 2 different variables.

4) Create a term with 3 different variables, 3 different exponents, and an irrational coefficient.

5) Create an expression with 2 terms, 2 different variables, and at least one irrational coefficient.

6) Create any equation in one variable that has the number 1 as the solution.

7) Create any equation in one variable, with at least 3 terms, that has the number 1 as its solution.

8) Create any equation in one variable, with at least 3 terms, that has 0 as its solution.

9) Explain why the word *isolate* is a synonym for solve.

10) In a few sentences, explain what it means to *solve an equation*.

11) In a few sentences, explain what it means to *simplify an expression*.

12) The acronym CLT means?

13) Given an equation and a solution, explain how you would check to see if the solution is correct.

14) State which of the following solutions (if any) makes the equation $-4x + 3 = 2 - x$ true.
   (a) $x = -1/3$ (b) $x = 1/3$
   (c) $x = 1$ (d) $x = 0$

15) State which of the following solutions (if any) makes the equation $5x + 1 = 3 - x - 2 + 6x$ true.
   (a) $x = 3$ (b) $x = 2$
   (c) $x = -5$ (d) $x = \pi$

16) State which of the following solutions (if any) makes the equation $8x - 2 = 4 + 8x$ true.
   (a) $x = 0$ (b) $x = -1$
   (c) $x = 1$ (d) $x = 10$

For questions 17 – 20, use the equation $y^2 - 1 = 2x - 7$

17) Check to see if x = 3, y = 0 is a solution.

18) Given that y = 3, find the value of x that satisfies the equation.

19) Without doing any work, if x = 3 what value of y will satisfy the equation? Why?

20) Check to see if x = 4, y = 2 is a solution.

21) Restate the *golden* and *silver* principles in your own words.

22) Given the following solved equation, write what was done for the golden and/or silver principles in each step. Step 1 has been done for you.

	6x – 10 = 3 + 7x	Silver principle	Golden principle
Step 1:	6x – 10 = 3 + 7x +10 +10 6x = 13 + 7x	10 was added to -10 to cancel it	10 was added to both sides
Step 2:	6x = 13 + 7x -7x -7x -x = 13		
Step 3:	-x = 13 (-1)-x = 13(-1) x = -13		

23) Explain what is incorrect in the following attempt at solving the equation…$3y + 2 = 15$

$$\frac{3y + 2}{3} = \frac{15}{3}$$

$\begin{array}{r} y + 2 = 5 \\ \underline{-2 \quad -2} \end{array}$

$y = 3$…………….CHECK: $3(3) + 2 = 15…9 + 2 = 15…11 = 15$..huh??

24) Given the equation $15x – 3.2 = 7x +1 – 2x – 5$, write-out your step-by-step solution in English and math. Step 1 has been done for you.

Step (1)   **English**: Combine like terms on the right-hand side to get one x pile and one number pile.

Math: $15x – 3.2 = 5x – 4$

25) For the following problems write the corresponding math or English phrase, whichever is missing. The first problem has been completed for you. Answers may vary.

Math	English
$3x – 10$	Three times a number (x) decreased by 10
	The sum of twice a number (x) and one-half of the same number
$\dfrac{2x}{7}$	

	The quotient of a number (x) and six, decreased by a different number (y)
$2(5x + 12)$	
$n + (n + 2) + (n + 4)$	
	The product of three consecutive, odd integers (use only the letter n)

26) Simplify the following expressions by combining like terms (CLT). Remember that like terms have the same structure; that is, the same variables and the same exponents.

((Examples of like terms include $5x^2y^3$ and $-3x^2y^3$. Also, $2xyz^4$ and $10xyz^4$ are like terms. The following terms are not alike: x^2y and xy^2))

(a) $6x + 4y - 12 + x - y + 7$

(b) $x^2 - y + 5 - 3x + 6y - 10$

(c) $3x - 4y + 5x^2y - 6yx^2 + 10xy - 3x^2y^2$

(d) $NPQ + PQN + QNP$

(e) $4(3 - 2x) + 6y(x - 1) - 5xy + 7$

(f) $2 - x + y - 2 - x + y - 2 + x - y$

(g) $7.01p - 5 + 8.9q + \pi - 4.2p + e + x$

(h) $5 - 3(2x - 1) + 6x + 10 - 2x(1 + x)$

27) I solved the equation for you below, but our illustrious illustrator forgot to write in some of the steps!! You'll have to do it for him. Fill in the missing algebraic operations in the boxes below.

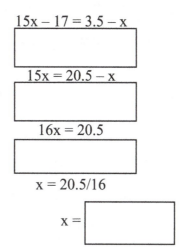

$$15x - 17 = 3.5 - x$$

$$15x = 20.5 - x$$

$$16x = 20.5$$

$$x = 20.5/16$$

$$x = \boxed{}$$

28) Solve the equation below two different ways and compare the number of steps and the level of difficulty for each solution... $12.6z - 18 = -6$

Isolate z by first removing the **Klingon**

Isolate z by first removing the **Dangler**

Compare the relative ease of the two approaches:

29) There is a local artist show in your town's library. You recognize some of the artists as up-and-coming, so you decide to attend and purchase a few pieces as investments. You have exactly $487 and you plan on spending all of it, if you can. It costs $35 to get in and each piece of art is priced at $75. Write an equation and solve for the number of art pieces you can buy? How much money do you have left over? Hint: Use the Modified Pólya Approach.

30) You are at the gas station filling up your gas tank. Show, using dimensional analysis, that dollars per gallon multiplied by gallons is dollars.

31) The acceleration of an object is the change in velocity over time…in other words velocity divided by time. Show, using dimensional analysis, that the unit for acceleration is meters per second-squared (m/s^2).

32) Solve the equations for the given variable. Remember to combine like terms when possible…keep the Klingon/Dangler principle in mind as well.

(a) $\dfrac{P}{4} = -12$

(b) $\dfrac{4}{P} = -7$

(c) $\dfrac{7}{x} = \dfrac{3}{x+1}$

(d) $6 - 2x = 5x + 9$

(e) $5(x - 1) = 95$

(f) $-3 = 3w - (4w + 6)$

(g) $19.6 - 21.3t = 80.1 - 9.2t$

(h) $-12 = 3(2x - 8)$

(i) $4y + 3(2 - y) = 5y + 10 - 2(1 + 2y)$

(j) $8(B + 2) = \frac{1}{2}(16B + 32)$

(k) $\frac{1}{5}(3T + 4) = \frac{1}{3}(2T - 8)$

(l) $\frac{x}{2} + \frac{x}{3} = \frac{1}{3}(x - 4)$

(m) $2T + 5(3 - T) - 7 = 6T(1 - 3 + 5) - T + 1$

(n) $\dfrac{2x}{3} = -16$

33) In each of the TI-83 graphs below, estimate the solution graphically. Each tick mark is 1 unit on the x and y axes.

(a)

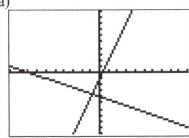

Estimate: (x, y) = _____

(b)

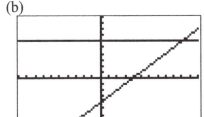

Estimate: (x, y) = _____

(c)

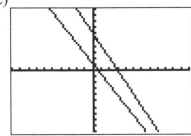

Estimate: (x, y) = _____

(d)

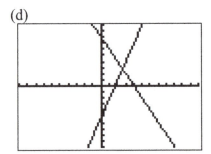

Estimate: (x, y) = _____

For questions 34 – 36, use the TI-83 to draw the graphs of the left and right sides of the equation (sketch them in the space provided). Then solve the equation using the Intersect method (6 magic buttons) on the TI-83. Keep in mind that you may need to play around with the window. Do your calculator solutions make sense given your sketches?

34) $-2x + 12 = x - 5$

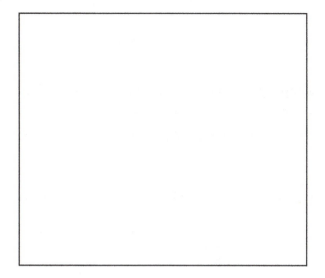

35) $6x + 20 = 12 - 2x$

36) $x - 5 = 3 + x$

Booklet on the Function Concept

What is a function?

Remember watching *Schoolhouse Rock* on Saturday morning? The conductor would always sing, "Conjunction junction what's your <u>function</u>?" Or maybe your grandma told you, "Now sweetie, your grades in school are a function of how hard you work!"…or maybe not. The word function in English usually means how something works or operates. What's the function of this little red butt – NO – don't touch that!......sorry.

In math, the word function has a very specific meaning, but it does relate to the English usage as well.

Definition of a function

A function is a special relationship between two variables (an input variable and an output variable). Every value of the input variable has one and only one value for the output variable.

Exactly one. No more, no less (as Yoda would say). Or, if you're a *Highlander* fan… "There can be only one!" Each x has only one y associated with it. One input, one output. Every horizontal point has only one corresponding vertical point.

A lovely non-example…One of our most beloved and used shapes of all time is not a function. That's right, the circle. A circle has two outputs for almost all of the inputs. ((by the way, all it takes is one instance of nonconformity to fail being a function))

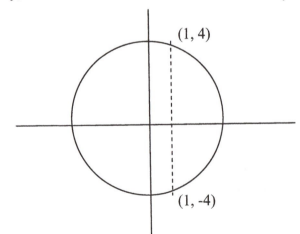

If you look at the input (x) value of 1, you will see 2 output (y) values associated with it…both 4 and -4. In fact, if you draw any **vertical line**, you will notice that it goes through two points, not one. There can be only one! A circle is not a function. It is a relationship between two variables, but that relationship is not a function.

<u>The Vertical Line Test (VLT)</u> is a good way to test if graphs are functions. If every vertical line goes through only one point, then the graph is a function. Consider the line below…

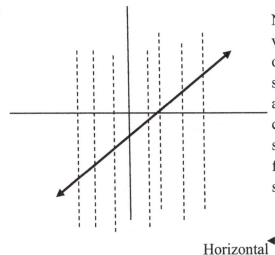

Notice that every vertical line drawn, and every vertical line that you could possibly draw will only go through one y-value (output). The VLT shows that this line is a function. In fact, all lines are functions except vertical lines. A vertical line contains infinitely many output values (not just one) so it cannot be a function. Horizontal lines are functions. Even though every output value is the same, each input has only one output. See below...

Horizontal line

Examples of functions:

(1) A candy vending machine represents a function. The two variables are amount of money put in and type of candy received. If you put in $1.25, will you get a bag of chips, a pack of gum, *and* a candy bar? No. If you said yes, please direct me to this vending machine!! One amount of money goes in, and one item comes out.

(2) An ATM is an example of a function. (P.S. Don't say ATM machine because the M in "ATM" already stands for machine). You put your information in (input) and you get out only your information (or money). You don't get your money and your neighbor's money, right?

(3) The relationship between decimals and fractions is a function. For every fraction, you have one decimal that represents it. We say that decimal representations are a function of fractional representations. If I give you 1/2, you give me 0.5. If I give you 2/5, you give me 0.4.

But what if we switch that around? Are fractional representations a function of decimal representations? Most certainly not! If I give you 0.5, you could give me 1/2, 2/4, 10/20, 300/600, etc. In this case, one input has way more than one output.

(4) The set of points {(0,2), (-3,6), (4, 12), (10, 0), (π, 5)} is a function because each input has exactly one output. In other words, each x has exactly one y.

But what if we changes the last point... {(0,2), (-3,6), (4, 12), (10, 0), (4, 5)}. This would not be a function because the input 4 has two outputs 12 and 5. If you were to plot these points on the x-y axes and do the VLT, it would fail at x = 4 because that vertical line would go through both 12 and 5.

You already know what functions are. In fact you have been using them your entire math career – just without knowing it. Functions have their own math *slang*, if you will, just like people from New Orleans. They say 'make and save groceries' instead of 'go grocery shopping and put the groceries away.' It's still English, but a slightly different usage. Function language is still Math language, just a slightly different usage.

In regular math language we say...   In function language we say...

In regular math language we say...	In function language we say...
x	input (or horizontal)
y	output (or vertical)
y = 2x	f(x) = 2x
depends upon	is a function of

Let's start with an English sentence and change it step by step to show how it becomes a function sentence...

"Your weekly pay is a function of how many hours you work."

The input variable is how many hours you work and the output variable is weekly pay. Notice that <u>output is always a function of input</u>...*muy importante!!*

Mathematicians are lazy. Instead of saying groups of words, we like to give them variable names. Let *P* be weekly pay and *h* stand for hours worked. So then we have...

P is a function of h

We know that *is* means =, so now we have...

P = a function of h

Finally, instead of saying "a function of" we write the letter f with parentheses...

P = f(h)

The way we read this is "P equals f of h." The parentheses don't mean multiplication in this case – they are just a symbol. In fact I call this the *symbolic* statement. All it does is tell you that there is a functional relationship, but it doesn't say what it is specifically. If we knew that you earned $10 per hour we could write the *specific* statement as well - $10 times *h* would be your weekly pay...

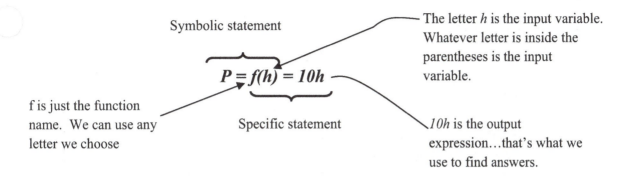

Symbolic statement

$P = f(h) = 10h$

The letter *h* is the input variable. Whatever letter is inside the parentheses is the input variable.

f is just the function name. We can use any letter we choose

Specific statement

10h is the output expression…that's what we use to find answers.

Psst!!…Is weekly pay truly a function of hours worked? Yes. If you work 25 hours you will get $250, not $250 *and* $120. Every amount of hours has exactly one amount of pay associated with it.

Example: Given the symbolic statement G = h(z), we know that z is the input variable, h is the function name, and G is the output. Make up an English statement that pertains to this function equation and write it on the line below.

_____

If we add a specific statement, like h(z) = $3z^2 - 5z + 4$, we can now actually crunch Captains…Uhmm…I mean…crunch numbers. If we want to find the output value when the input value is 1, we write h(1) = $3(1)^2 - 5(1) + 4 = 3 - 5 + 4 = 2$.

Instead of actually saying, "Find the output value when the input value is 1" all we have to write is "h(1)". Whatever takes the place of the z in parentheses on the left hand side of the equation gets substituted for every instance of z on the right hand side of the equation as well.

h(z) = $3z^2 - 5z + 4$
h(0) = $3(0)^2 - 5(0) + 4$
h(-6) = $3(-6)^2 - 5(-6) + 4$
h(100) = $3(100)^2 - 5(100) + 4$
h(π) = $3(\pi)^2 - 5(\pi) + 4$
h(blue) = $3(blue)^2 - 5(blue) + 4$
h(Ashley) = $3(Ashley)^2 - 5(Ashley) + 4$
h(?) = $3(?)^2 - 5(?) + 4$

Example: *The square of a number is a function of the number*. Let's call the output variable y, the input variable x, and the function name f. I will do the following: (a) State whether or not this relationship is a function, (b) write the symbolic statement, (c) write the specific statement, and (d) find the output values for inputs of 4, 5, and 9.

(a) This statement is a function because for every number there is only one square. For example, the square of 2 is 4 (not 4 and 8 and -13).

(b) The symbolic statement is $y = f(x)$. The output variable y (square of number) is a function of the input variable x (number).

(c) The specific statement is $f(x) = x^2$. The way you find the square of a number is by squaring the number…Am I repeating myself?? Putting them together you have $y = f(x) = x^2$. In English that says, "y is a function of x and specifically it is the value of x squared."

(d) $f(4) = 4^2 = 16$
        $f(5) = 5^2 = 25$
        $f(9) = 9^2 = 81$

These (input, output) pairs can also be written as (4, 16), (5, 25), and (9, 81). In function language, writing (x, y) is the same as writing (x, f(x)), (horizontal, vertical), or (input, output).

What if we swapped input and output variables in our example and made the statement, '*A number is a function of its square.*'? Would that still be a function? Now the input is the square and the output is the number. So if I give you 81 as a square, you would give me both 9 and -9 as the output because $\sqrt{81} = 9$ and $\sqrt{81} = -9$. Since each input has two outputs, this would not be a function.

✿ *In general*, If A is a function of B, then B is *not necessarily* a function of A.

Write two of your own examples of this concept. For each example, think of two variables that create a functional relationship one way, but *do not* when the input and output variables are swapped.

Example 1:
Example 1 (swapped):
Example 2:
Example 2 (swapped):

Professor's Practice Pause

A. State your definition of a function. Later on, ask your dorm RA what his/her definition is and compare the two.

B. BLT stands for Bacon, Lettuce, and Tomato. What does VLT stand for and what do you use it for?

Γ. State whether the following sentences are true or false and explain your answers. Remember to use the definition of a function.

  (a) Height of people is a function of their weight.

  (b) Weight of people is a function of their height.

  (c) Area of a circle is a function of radius.

  (d) Cost of filling your gas tank is a function of number of gallons pumped.

Δ. B = {(0, 5), (3, 5), (-12.6, 5), (-5, 5), (100, 5)}. Does this set represent a function? Explain.

Warning – big word alert

E. The pressure of a gas inside a sealed container is a function of the *ambient* temperature. In this instance, the pressure is 70 degrees less than five times the temperature. Write <u>symbolic</u> and <u>specific</u> function statements for the above information. Use P for pressure, T for temperature, and G for the function name.

Z. Given $f(x) = -2x^3 + 2x - 5$, find $f(0)$, $f(-1)$, and $f(1)$.

Associated End of Booklet exercises are 1 – 20.

What do we mean by "Families" of functions?

La Famiglia. Der Familie. The family. What is this whole business about families of functions, anyway? What do we mean by that? Well…let's look at our families. They are typically made up of similar-looking and similarly-behaving people. So much so that oftentimes it's hard to tell one sibling from another, or difficult to distinguish mother from daughter. Every girl knows that she'll eventually turn into her mother. The same is true regarding families of functions: they have the same basic shape and the same basic properties, but they are not exact replicas of one another. Mathematically, these differences are generated by taking a *basic* member of the family and performing a *translation*.

A translation is simply a shift up, down, left, or right on the x-y axes by a constant value. The translated members of a family have the same characteristics, same shape, and similar-looking equations, but are still not exactly the same function. Families of functions become important in calculus, differential equations, geometry, engineering, economics, and other physical sciences. In fact, nothing that you never learn in this book will ever be unimportant, ok!?

Let's look at some typical families – just enough to get a basic understanding. We will study each family in more detail in different Booklets.

Example: The colored lines shown below are all members of the linear family of functions with the equation $L(x) = x + a$, where a (in this example) is some integer. For example, $L(x) = x$, $x + 1$, $x + 2$, $x - 1$, $x - 2$, etc. Notice they have the same shape, but are just moved up and down the y-axis. The graph on the right shows the same family members on the TI-83.

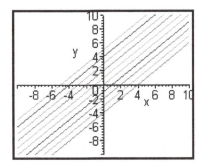  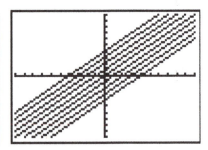

Example: The graph below represents members of the family of quadratic functions, all with the equation $Q(x) = (x^2 + a) + b$, where a and b are integers. Notice the same similarities as the linear family, although now we see movement, not only up and down on the y-axis, but also left and right on the x-axis.

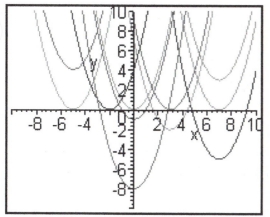

The specific functions graphed on the left are $Q(x) = x^2$

$(x-7)^2 + 3$

$(x+2)^2$

$(x+5)^2$ $x^2 - 1$

$(x-3)^2$

$(x-7)^2$ $(x-3)^2 - 2$

$x^2 - 8$

$(x+5)^2 - 4$

$(x-7)^2 - 5$

The question is, how do we create these other family members using translations? Ahhh, but I am about to show you…

Starting with any basic function, f(x), there are 5 changes that are important to us: f(x) + a, f(x) − a, f(x + a), f(x − a), and −f(x). Let's see what they do individually.

(1) f(x) + a raises all points on the function by a units. So f(x + a) is a shift up.

(2) f(x) − a lowers all points on the function by a units. So f(x − a) is a shift down.

Here are the two tricky ones…

(3) f(x + a) moves every point on the function to the <u>left</u> by a units. *Why to the left?* Look at the equation x + a = 0. If a = 2, then solving the equation for x gives x = -2. The x moved to the left when a is a positive value.

(4) f(x − a) moves every point on the function to the <u>right</u> by a units. *Why to the right?* Look at the equation x − a = 0. If a = 2, then we have x − 2 = 0…solving the equation for x gives x = +2. The x moved to the right when a is a negative value.

(5) −f(x) is really called a **reflection**, not a translation, because it "reflects" the graph of f(x) over the x-axis. If every value of a graph were to become its negative, then the entire graph would be flipped (or reflected) over the x-axis.

TRANSLATION SUMMARY	
When f(x) goes to…	The graph shifts…
f(x) + a	UP by a units
f(x) − a	DOWN by a units

f(x + a)	LEFT by a units
f(x − a)	RIGHT by a units
-f(x)	OVER the x-axis

We can also combine the translations to form members that have been moved not only right and left, but up and down as well. The function f(x + a) + b accomplishes this.

Example: Given the function f(x) = x + 4, does f(x + 2) move the entire function up by 2 units?
f(x + 2) = (x + 2) + 4 = x + 2 + 4 = x + 6…yes.

Example: Given the function f(x) = x². The graph of this function has its lowest point at (0,0).

(a) What is the lowest point of f(x − 3)? Since this translation moves the function to the right by 3 units, the new lowest point will be (3,0).

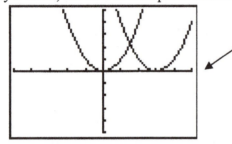
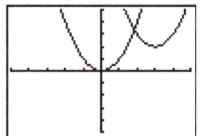

(b) What is the lowest point of f(x − 3) + 2? Since this translation moves the function to the right by 3 units and up by 2 units, the new lowest point will be (3,2).

Example: Given the function g(x) = (1.5)ˣ, the new member of the family created by g(x + 4) − 5 should slide g(x) down 5 units and left 4 units. While it might not be so easy to see the translation from the graph below, we can check algebraically, too.

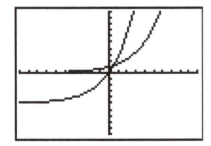

Algebraic check

Pick an x-value of 1. g(1) = 1.5¹ = 1.5, so we have a point on g(x)…(1, 1.5).

We know that the translation should take the point (1, 1.5) and send it left 4 and down 5 units. That would make the point (1 − 4, 1.5 − 5) = (-3, -3.5).

When x = -3 on the translated function, g(x+4) − 5 becomes 1.5^(-3 + 4) − 5 = 1.5¹ − 5 which equals -3.5, so we have the point (-3, -3.5)…the translation checks!

Professor's Practice Pause

Explain in **_words_** what the following translations will do and verify them on your TI. The starting function is on the left, and the translation is after the semi-colon.

(a) $f(x) = x - 7$; $f(x - 3)$

(b) $g(x) = x^3$; $g(x - 1) + 2$

(c) $h(x) = x^2$; $h(x + 3) - 4$

(d) $f(x) = 3 \cdot 2^x$; $f(x + 4)$

Associated End of Booklet exercises are 21 – 24.

Algebra of functions (+, -, x, but not ÷)

If someone happens to give you some functions for Christmas, what will you do with them? Yes, I know – you're thinking, 'Why in the world would you give functions for Christmas?' Well, kids really love 'em…ask mine! Plus you don't need batteries, they never get old, and they afford hours upon hours of enjoyment. There's always that difficult person on your Christmas list for whom you never know what to buy. Next time, try writing some functions down on strips of paper (maybe wrapping them individually for that special touch), sit back, and watch the expressions of sheer joy as they are opened. As Henry David Thoreau stated so eloquently:

"A noble person confers no such gift as his whole confidence: none so exalts the giver and the receiver; it produces the truest gratitude. Perhaps it is only essential to friendship that some vital trust should have been reposed by the one in the other. I feel addressed and probed even to the remotest parts of my being when one nobly shows, even in trivial things, an implicit faith in me.... A threat or a curse may be forgotten, but this mild trust translates me."

P.S. This quote really has nothing to do with giving functions as gifts…I just like it!

Seriously, though, what could you do with functions if you *were* given them? Well, the same things that you could do with numbers, I guess – add, subtract, multiply, and divide. We'll focus our attention on the first 3 operations and leave division for the another course.

Adding and subtracting functions is easy…all you do is CLT (combine like terms). Now we have BLT, VLT, and CLT as acronyms. On the line below, state what they mean in mathematics, and then make up your own acronym that has anything to do with the current state of the economy.

ACRONYM EXERCISE

	real meaning	make your own
BLT	_____	_____
CLT	_____	_____
VLT	_____	_____

Example of adding functions: Given $f(x) = 5x^2 - 3x + 2$ and $g(x) = 12x - 7$.

$$f(x) + g(x) = 5x^2 - 3x + 2 + (12x - 7) = 5x^2 \underbrace{(-3x + 12x)}_{CLT} + \underbrace{(2 - 7)}_{CLT} =$$

$$f(x) + g(x) = 5x^2 + 9x - 5$$

Example of adding functions: Given the same functions as above, find $f(x) - g(x)$

$$f(x) - g(x) = 5x^2 - 3x + 2 - (12x - 7) = 5x^2 - 3x + 2 - 12x + 7 \quad \boxed{\text{Remember the Distributive law}}$$

$$= 5x^2 \underbrace{(-3x - 12x)}_{\text{CLT}} + \underbrace{(2 + 7)}_{\text{CLT}}$$

$$f(x) - g(x) = 5x^2 - 15x + 9$$

Notice that $f(x) + g(x)$ is not equal to $f(x) - g(x)$. We expected that! Functions behave like numbers with respect to these properties…

$$f(x) + g(x) = g(x) + f(x) \ldots \text{commutative law of addition}$$
$$f(x)g(x) = g(x)f(x) \ldots \text{commutative law of multiplication}$$

$$f(x) - g(x) \neq g(x) - f(x) \ldots \text{subtraction does not commute (in general)}$$
$$f(x)/g(x) \neq g(x)/f(x) \ldots \text{division does not commute (in general)}$$

Professor's Practice Pause

Perform the operations below, given the following functions:
$$G(t) = -2t^2 + 4t - 9, \ F(t) = 3 - 12t, \text{ and } P(t) = t^2 + 5$$

(a) $3G(t) - P(t)$

(b) $F(t) + G(t)$

(c) $P(t)/2 - F(t)/3$

(d) $G(t) - F(t) + P(t)$

Associated End of Booklet exercises are 25 – 28.

Addition and subtraction of functions – *Group Pause*

Renting a mini-van on your Spring break trip to sunny Florida

You are renting a mini-van for a week with some college friends. The van costs $345 for the week, plus $0.03 per mile. By checking the AAA website, you can expect to pay an average of $3.18 per gallon of gas along the way. The rental agency tells you the van gets 25 mpg.

(a) Rental cost is a function of miles driven. R = h(n). Write a specific function for the rental portion of the cost (R) as a function of the number of miles driven (n). R = h(n) = ?

(b) Fuel cost is also a function of miles driven. F = g(n). Write a specific function for the fuel portion of the cost (F) as a function of the number of miles driven (n). F = g(n) = ?

(c) Write a specific function for the total cost (C) of driving the van as a function of the number of miles driven (n). C(n) = h(n) + g(n). Simplify your answer.

(d) If you drive 2500 miles for the week, what is your total cost?

(e) Explain the practical meaning of the math statement C(1750) = 620.10.

(f) Another rental agency had a mini-van for $500 a week, but with unlimited mileage. Write the total cost function for driving this mini-van. If you knew you had to drive 3000 miles (to Florida and back), which mini-van would have been cheaper to rent from the beginning?

Multiplication of functions

I would like to show you 2 different ways to multiply functions together (specifically linear functions and polynomials). One method is called the FOIL method which stands for "First Outer Inner Last." You've no doubt heard of this method and used it before. If not, you certainly ran into it during the Acronym Exercise in the Booklet on Complex Numbers. Do you remember the acronym that you created for it? The second method is called the Box method, which I like better because it's more organized – students need organization skills in math…like bow-staff skills, nunchuck skills, etc.

The FOIL method – pronounced (foy-ul) unless you're from Texas in which case it's pronounced (ferl)

The FOIL method is really set up for multiplying two binomials together (a binomial has two terms. The prefix "bi" means two – like bicentennial and bicycle. An example is the best way to show it…

$$(2x - 4)(3x + 1) = (2x)(3x)\text{…the \textbf{First} terms in each binomial}$$
$$+ (2x)(1)\text{…the \textbf{Outer} terms in each binomial}$$
$$+ (-4)(3x)\text{…the \textbf{Inner} terms in each binomial}$$
$$+ (-4)(1)\text{…the \textbf{Last} terms in each binomial}$$

$$(2x - 4)(3x + 1) = (2x)(3x) + (2x)(1) + (-4)(3x) + (-4)(1) = 6x^2 + 2x - 12x - 4$$
$$= 6x^2 - 10x - 4$$

Schematic of FOIL

$$(a + b)(c + d)$$

with arcs labeled I, L, F, O

The Box method

The FOIL method is good. You know – it's been around for awhile, all of your teachers love to teach it (including the indomitable Mrs. Bumblemeyer). But what happens when the polynomials get larger…like $(x^2 + 4x - 3)(-2x^2 - 7x + 9)$? If you use FOIL for this, you will end up with 9 different terms all in a row!! Very confusing!! The box method will help you stay organized and give you a clue on how to combine your answer at the end. As Napolean Dynamite would say, "Sweet!"

Example: Multiply $(x^2 + 4x - 3)$ and $(-2x^2 - 7x + 9)$ using the box method.

Count the number of terms in each polynomial and create a box with those dimensions. Since there are 3 terms in both, I am going to make a 3 x 3 box. I will also place one of the polynomials along the top and one along one side…like this…

	x^2	$4x$	-3
$-2x^2$			
$-7x$			
9			

Now…we complete the 9 boxes by **multiplying the row and column** of that box. For example, the center square would be $4x(-7x) = -28x^2$.

	x^2	$4x$	-3
$-2x^2$	$-2x^4$	$-8x^3$	$6x^2$
$-7x$	$-7x^3$	$-28x^2$	$21x$
9	$9x^2$	$36x$	-27

Now we want to add them up (combine like terms). Notice, however, that this is made easier because boxes along the diagonals have the same exponent.

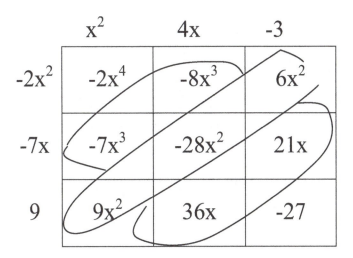

	x^2	$4x$	-3
$-2x^2$	$-2x^4$	$-8x^3$	$6x^2$
$-7x$	$-7x^3$	$-28x^2$	$21x$
9	$9x^2$	$36x$	-27

Adding up diagonals, we have $-2x^4 - 15x^3 - 13x^2 + 57x - 27$

Example:

If you completed the research problem about the Mona Lisa in the Booklet on Real Numbers, you know that the original painting has a height of 77cm and a width of 53cm. Imagine larger prints are authorized to be made from the original that increased the dimensions of height and width by a certain amount (x), with the further stipulation that the added height must always be twice as long as the added width.

The new size of the print would be as follows:

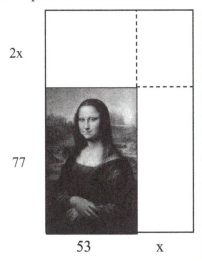

We now have a new width, $x + 53$ and a new height, $77 + 2x$

The original area of the Mona Lisa was $53(77)$ or 4081 cm^2. Using FOIL, the new area of the larger print is

$$A(x) = (x + 53)(2x + 77)$$
$$= 2x^2 + 77x + 2(53)x + (53)(77)$$
$$= 2x^2 + 77x + 106x + 4081$$
$$A(x) = 2x^2 + 183x + 4081$$

Notice what happens below if we use the **Box** method.

	53	x
2x	$A = 53(2x)$ $= \underline{106x}$	$A = x(2x) = \underline{2x^2}$
77	$A = (53)(77)$ $= \underline{4081}$ (Original Photo)	$A = \underline{77x}$

If you add up the area of all 4 boxes, you get the same area…

$$A(x) = 2x^2 + 77x + 106x + 4081$$
$$A(x) = 2x^2 + 183x + 4081$$

The Box method is just a graphical representation of the FOIL method

Question: Given the area function above, find the area of the print when 6cm is added to the width. In other words find A(6).

$$A(6) = 2(6)^2 + 183(6) + 4081 = 2(36) + 183(6) + 4081 = 72 + 1098 + 4081 = 5251 cm^2.$$

Professor's Practice Pause

Given the functions $f(x) = 5x^2 - 4x + 12$ and $g(x) = 7x - 8$, find $f(x)g(x)$ using both the FOIL and Box methods. When you are done, write which one you like better and why.

Associated End of Booklet exercises are 29 – 32.

Math Modeling with functions

"Math Modeling" might conjure images of scantily clad numbers parading on the catwalk, but that's not really what it means. It simply means finding equations that fit the data. Later on we'll call this *regression* (which has nothing to do with hypnosis). But I digress…Once we have a suitable equation (model) for some data, we can then make predictions and see what will happen in the future (*extrapolation*), or see what happens in between some missing data points (*interpolation*).

Language note...

The prefix *extra* means outside, beyond, or besides as in extraterrestrial, extraordinary, or extrapolate (outside the data).

The prefix *inter* means between or among as in interstate highway, intercontinental, or interpolate (between the data).

The mechanical robot that roved on Mars several years ago recorded all sorts of geographical and natural data about the planet. Upon re-entry and further analysis of the rover, NASA scientists discovered that the robot carried back with it an unknown strain of "bacteria." The growth of this bacteria was measured over a period of a few days. It was discovered that the amount of bacteria is a function of time. Some of the data is recorded below (Source: Data stolen from AREA 51 in Roswell, New Mexico)

Time (hours)	0	1	2	3	4	5	6
Bacteria	2500	3500	4900	6860	9604	13446	18824

Creating a scatter plot on the TI-83

Data Entering Mode

1. Press the **STAT** button
2. Press **1** or **ENTER** (to get the EDIT mode)

3. Enter your x (input) data into L_1, and your y (output) data into L_2

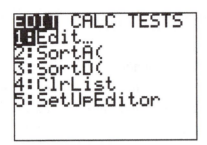

** Be sure there are the same amount of entries in each list. Don't worry if there is data in **L₃**.

If you ever want to clear a list completely, place the cursor on the list name (**L1**, for example), then press **CLEAR** and **ENTER**.

Scatter Plot Mode

4. Press the **2nd** and **Y=** (to get the STATPLOT mode)
5. Press **1** or **ENTER** (to get into PLOT 1)

6. Highlight the following on the Menu...
      ON
      TYPE (**first one**…scatter)
      XLIST: **L1**, or whatever list you put the x data in
      YLIST: **L2**, or whatever list you put the y data in
      MARK: Pick one (it doesn't matter)

NOTE: In order to select the lists, press **2ⁿᵈ** and then either **1, 2, 3, 4, 5,** or **6**. Do _not_ type in the letter "L" and then a number, you must use the L's given on the TI.

7. Press the **WINDOW** button. Make sure your window _covers_ the data. Since our input data goes from 0 to 6, I'll choose an x window of -1 to 7 (a little less than the lowest data value and a little more than the highest data value). For the output (y) data, I will choose a window from 2000 to 20000. Choosing a window is definitely more of an art than a science. You should always be playing with the window until you get a picture that you like.

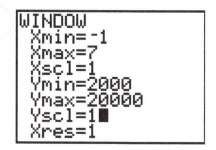

8. Press **Y=**…and clear unwanted equations. Notice that **Plot1** is highlighted…that's good.

9. Press **GRAPH**. You should see the scatter plot below…If you do not see this scatter plot, first go back to step 1 and make sure everything looks like my TI pictures. If that doesn't work, stand on your table and scream as loud as you can…No, no – just kidding! Raise your hand and ask your professor to help you, or, if you are in groups, ask your mathematical neighbor. Your math neighbor knows everything!

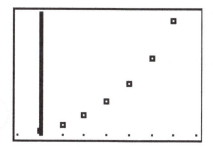

Now, we can answer some questions…

A. Given the following mathematical models (functions), determine which best represents the data:

(a) f(x) = 250x + 2400   (b) h(x) = $3x^2$ + x + 2000

(c) g(x) = 2300(1.4)^x (d) k(x) = 2500/x

In order for you to figure this out, plot each function (in **Y1**) one at a time (or maybe all at once), and see which model best fits the data. I have done letter (a) for you below…

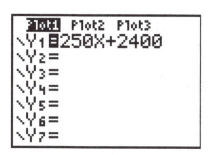

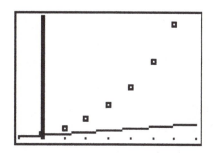

Clearly, the function f(x) = 250x + 2400 does not match the data that well, right? Try the other three and write the equation of the function that best models this data below...

BEST MODEL _____

B. Using your chosen "best" model, write a *symbolic* function sentence showing the output when the input is 7. This is an example of extrapolation (making predictions outside the data).

Table Features

To use the Table features of the TI, you must first enter an equation into **Y1**. For example, function (a) above...Y1 = 250x + 2400. Then press **2nd** and **WINDOW**.

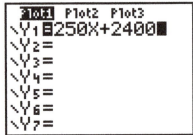

There are two choices for you...

Choice 1 – You can enter the x data and the TI will give you the y data automatically. This is done by selecting **Indpnt:Ask** (as shown above)

Choice 2 – You can have the TI automatically show x and y data. To do this, select **Indpnt:Auto.** You will need to pick a starting value (**TblStart**), and an incrementing value (**ΔTbl**).

Below are two examples, the one on the left shows the Auto option, and the one on the right shows me entering 3 input values (0, 2, and 25).

Back to the bacteria questions…

C. Using your chosen "best" model and the Table features of the TI, predict the growth of the bacteria after 8 hours. Fill in the answer below.

f(8) = _____

Professor's Practice Pause

Modeling with Functions - Group Exercise

A rocket launcher mounted on the deck of a US naval destroyer fires a rocket at an approaching enemy fighter plane. The launching unit is 30' above sea level. The table below represents the height of the rocket above sea level as a function of time.

Time (seconds)	Height above Sea level (feet)
0	30
2	45
3	67
5	125
7	180
8	248
12	420
15	496

A. Create a scatter plot of your data.

B. Of the four mathematical models below, which one best models the data:

(a) $f(x) = 40x - 2$

(b) $f(x) = -0.22x^3 + 5.8x^2 - 6.3x + 33.6$

(c) $f(x) = -0.9x^2 + 21x + 10$

(d) $f(x) = 100x - 25$

C. Explain in a few sentences how you found your answer to the question B?

D. Using your answer to question B, how high should the rocket be after 13 seconds? Note that this is an example of <u>interpolation</u> (filling in the missing data).

Associated End of Booklet exercises are 33 – 36.

Domain and Range of functions

Every function has a *domain* and a *range*. Unfortunately, the math definitions are not really associated with the English definitions for these words. Many students have trouble with these two concepts, but I'm about to break it on down for ya…it's like – How do you know where I'm goin' if you aint been where I been? Understand where I'm coming from?

Definition of Domain: The set of all allowable (or reasonable) values that can be substituted for the input variable. The domain is a set of x-values.

Definition of Range: The set of all corresponding output values. The range is a set of y-values.

Another way to think about it… Since the x-values are on the horizontal axis, you could say that the domain goes from the *leftmost* x-value to the *rightmost* x-value. Since the y-values are on the vertical axis, you could say that the range goes from the *bottommost* y-value to the *topmost* y-value.

$$D_f = [\text{leftmost, rightmost}]$$
$$R_f = [\text{bottommost, topmost}]$$

The symbol "D_f" means the domain of the function, and "R_f" means the range of the function.

Math to English

Remember that the open parentheses, ( ) mean that the values do not include endpoints, whereas the closed brackets, [ ] do include endpoints.

The best way to understand these two concepts is to look at many examples. I have provided many examples for you to look at. Remember, only functions have a domain and range.

Examples:

1. $f(x) = \{(0,0), (3,-5), (7,10), (-1.75,22.68), (4,6)\}$

This is a function.
$D_f = \{0, 3, 7, -1.75, 4\}$
$R_f = \{0, -5, 10, 22.68, 6\}$

For sets of coordinate points, the domain is just a listing of all the x-values and the range is a listing of all the y values.

2. $f(x) = \{(0,0), (3,0), (7,0), (4,0)\}$

This is a function.
$D_f = \{0, 3, 7, 4\}$
$R_f = \{0\}$

3. $T(x) = \{(5,0), (5,-5), (5,10), (5,6)\}$

This is not a function. The input of 5 has multiple outputs, therefore the graph of this function does not pass the vertical line test.

4.

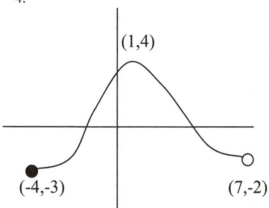

This is a function.
$D_f = -4 \leq x < 7$
$R_f = -3 \leq y \leq 4$

5.

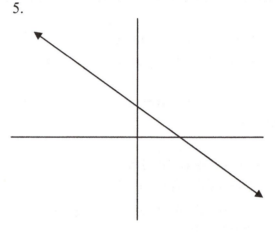

All lines with slope are functions. The D_f and R_f are all real numbers because there are no left-, right-, top- or bottommost values. Lines with slope continue in all 4 directions.

6.

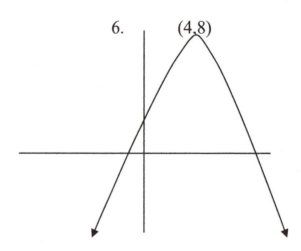

This is a function.
D_f = all real numbers (no leftmost or rightmost values)
$R_f = y \leq 8$ (8 is the topmost y-value).

In set notation we would write: $D_f = (-\infty, \infty)$ and $R_f = (-\infty, 8]$

'. This curve continues to the left, right, and up.

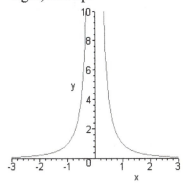

This is a function.
$D_f = x \neq 0$ (all x values except 0)
$R_f = y > 0$ (only positive y values)

8. This curve continues to the right, but begins at -6

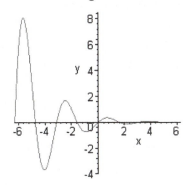

This is a function.
$D_f = x \geq -6$
$R_f = -4 \leq y \leq 8$

9. This curve continues to the left, right, and up.

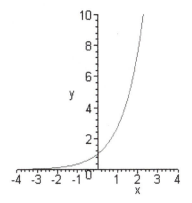

This is a function.
D_f = all real numbers
$R_f = y > 0$ (only positive y values)

10. This curve continues to the right and up, and begins at (-3,-5)

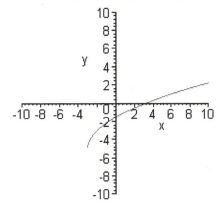

This is a function.
$D_f = x \geq -3$
$R_f = y \geq -5$

11. $f(x) = x^3$

This is a function.
D_f = all real numbers
R_f = all real numbers

12. $g(x) = 7$

This is a function.
D_f = all real numbers
$R_f = \{7\}$ (see example 2)

Composition of functions

We have tried to use music and art examples in these Booklets as much as possible to support *The Arts* whenever we could, and show the links between math, music and art. You would think that *composition* means something musical – but, it doesn't! Sorry. Here we have a perfect musical word that is purely mathematical.

The composition of two functions (composing functions) is an operation separate from the other. $f(x) \pm g(x)$ means adding or subtracting functions by combining like terms. $f(x)g(x)$ means multiplying two functions using either the FOIL or Box methods. Composing two functions is taking one function and making it the input to the other function – putting one inside the other. It looks like this $f(g(x))$, read "f of g of x." Let's go back to our original example of working with functions…$h(z) = 3z^2 - 5z + 4$

$$h(0) = 3(0)^2 - 5(0) + 4 \qquad \text{Just like these…}$$
$$h(-6) = 3(-6)^2 - 5(-6) + 4$$
$$h(100) = 3(100)^2 - 5(100) + 4$$
$$h(\pi) = 3(\pi)^2 - 5(\pi) + 4………………\text{and so on}$$

If we know introduce $g(z) = 10 - 8z$, and say $h(g(z))$, it means take $g(z)$ and put it into h..

$$h(g(z)) = g(z)^2 - 5g(z) + 4$$

But, we know what $g(z)$ is, don't we? $g(z) = 10 - 8z$, so we can simplify even more…

$$h(g(z)) = (10 - 8z))^2 - 5(10 - 8z) + 4$$

Notice the structure of h is still the same…$h(something) = 3(something)^2 - 5(something) + 4$. The only thing we were changing in the examples above is the *something*. In the case of the composition, $h(g(z))$, the something is the other function, $g(z)$.

Viewing composition as an input/output machine

Here's another illustration for understanding the composition $f(g(x))$

$$x \rightarrow \boxed{\quad g \quad} \rightarrow g(x) \qquad g(x) \rightarrow \boxed{\quad f \quad} \rightarrow f(g(x))$$

$$\text{IN} \qquad\qquad \text{OUT} \qquad\qquad \text{IN} \qquad\qquad \text{OUT}$$

x is the input to the function g. It comes out as g(x). Then g(x) is the input to the function f, which comes out as f(g(x))…"f of g of x". This could continue with many functions. We could string input and outputs together all day long to create f(g(h(z(p(x))))).

$$\text{Given } f(x) = x^2 - 2x + 1, \; g(x) = -5x - 9, \text{ and } h(x) = x^3 + 4x$$

(a) Find f(g(x)). f(g(x)) = f(-5x – 9). The function f takes an input and turns it into $(\text{input})^2 -$ 2(input) + 1, but we know what the new input is..-5x – 9. In other words "x" becomes "-5x – 9" So, f(g(x)) = $(-5x - 9)^2 - 2(-5x - 9) + 1$. That's a satisfactory answer. However, if we wanted to work with that function some more, we would want to simplify it and combine like terms…

$$\begin{aligned} f(g(x)) &= (-5x - 9)^2 - 2(-5x - 9) + 1 = (-5x - 9)(-5x - 9) - 2(-5x - 9) + 1 \\ &= (25x^2 + 45x + 45x + 81) + (10x + 18) + 1 \\ &= 25x^2 + 100x + 100 \end{aligned}$$

(b) Find g(f(x)). Now f is the input for the function g. Let me show you another way to view composition…g(f(x)) = -5(f(x)) – 9. Again, we know what f(x) is…$x^2 - 2x + 1$, so we have..

$$g(f(x)) = -5(f(x)) - 9 = -5(x^2 - 2x + 1) - 9.$$

The new input
is f(x)

Sehr wichtig Punkt! Translation…Very important point! Notice that the composition one way, f(g(x)), did not equal the composition the other way, g(f(x)). In general, composition does not commute. There is, however, one important situation when they do, but we will discuss that in just a few pages…

(c) Find h(g(1)). Notice now that we are finding an actual input value. There are 2 ways to do this. The first way is to find h(g(x)) and then substitute x = 1. The second way would be to substitute x = 1 into g(x) and then calculate h(g(x)). I will, of course, show you both ways --

<table>
<tr><th>Way #1</th><th>Way #2</th></tr>
</table>

$$h(g(x)) = g(x)^3 + 4 \cdot g(x)$$
$$= (-5x - 9)^3 + 4(-5x - 9)$$
$$= (-5x - 9)(-5x - 9)(-5x - 9) + 4(-5x - 9)$$
$$= (-125x^3 - 675x^2 - 1215x - 729) + (-20x - 36)$$
$$= -125x^3 - 675x^2 - 1235x - 765$$
$$h(g(1)) = -125(1)^3 - 675(1)^2 - 1235(1) - 765$$
$$h(g(1)) = -125 - 675 - 1235 - 765 = \mathbf{-2800}$$

$$g(1) = (-5(1) - 9) = -14$$
$$h(-14) = (-14)^3 + 4(-14)$$
$$= -2744 - 56 = \mathbf{-2800}$$

Hmmm…I wonder which
way is easier??

Hey, I forgot to show you **Way #3**…Type in h(g(x)) in Y_1 in your TI and use the Table method.

** Notice that the TI will do the composition for you. You do not have to simplify before entering in the calculator. In actuality, the TI does not *algebraically* expand h(g(x)) like you would by hand, it uses a numerical algorithm to simply find h(g(1))….stupid, but very fast!

Application Example: Oil spilled from an Exxon tanker will approximate the shape of a circle on the surface of the ocean. The area of the spill (in square meters) is a function of the radius of the oil circle, but the oil circle is a function of seconds (the more time, the larger the radius of the spill). We know that $A(r) = \pi r^2$, and we are told by EPA chemists over the phone that the radius of the oil circle will increase by 25 meters every hour. Oh, I forgot to tell you…you're the blasted Captain of that ship and you're going to be in big, big trouble! How many square meters of oil will be on the ocean's surface in 10 hours if the spill is not stopped?

We know that area is a function of radius…$A(r) = \pi r^2$

We know that radius is a function of hours $r(h) = 25h$ (25 times the number of hours)

We have A as a function of r, and r as a function of time in hours. We want A to be a function of hours, so we use composition to combine the variables and eliminate the r…

$A(r(h)) = \pi(r(h))^2 = \pi(25h)^2 = \pi \cdot 625h^2$. So, $A(h) = 1963.5 \cdot h^2$

Since we want to find the amount of oil in 10 hours, we need to find A(10)

$A(10) = 1963.5 \cdot (10)^2 = 196{,}349$ m^2 of oil.

If the oil spill is not stopped before the 10 hour mark there will be almost 200,000 square meters of oil on the surface of the ocean.

Inverse of a function

Whenever I ask students the meaning of the word *inverse*, a typical first response is "opposite". To some extent that's true, but then I usually say, 'ok, what's the opposite of 5?' Some students say -5 and other students say 1/5...and they are both correct. -5 is the <u>additive</u> inverse of 5 and 1/5 is the <u>multiplicative</u> inverse of 5. The inverse of a function is something different.

Graphically, the inverse of a function is a reflection of all points on the curve over the line y = x (what I call the *mirror* line). That means that a point (x, y) on the function will become (y, x) on the inverse.

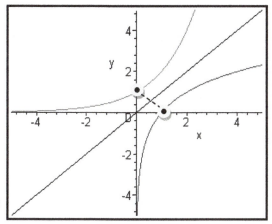
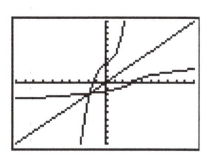

In the colored graph above, the top function and the bottom function are inverses of each other. They are flipped over the black (mirror) line (y = x). Notice that the point (0,1) on the red function is mapped perpendicularly across the mirror line to become (1,0). The graph on the right is an example of two inverse functions and the mirror line on the TI-83.

The **algebraic** definition of a function uses composition...I told you we would see that in just a few pages...in fact, I told you that just a few pages ago.

f(x) and g(x) are inverses iff...

$$f(g(x)) = g(f(x)) = x$$

In English, we would say, "two functions are inverses if the composition one way is the same as the composition the other way, and both of them equal the input."

Here's where we get into a bit of a fix! The notation that we use for "the inverse of a function" is $f^1(x)$. Whoever came up with that should be shot! This has confused college algebra students for centuries. The symbol is not math, it's just symbolism. $f^1(x)$ is read "f inverse of x." And, most importantly...

$$f^{-1}(x) \neq \frac{1}{f(x)}$$ That's where the confusion comes in because all

throughout your math history, the indomitable Mrs. Bumblemeyer told you that x^{-1} means $1/x$...and it does! – just not with functions. Sorry...I really do apologize for the tricky notation. You may have remembered the same problem in trigonometry (if you took trigonometry in high school). The inverse for the sine function ($\sin x$) is also written as $\sin^{-1}x$, but again that <u>does not</u> mean $1/\sin x$. A better notation for the inverse sine function is arcsin x, which you will see in certain texts. Perhaps instead of $f^{-1}(x)$, we should say $f_{arc}(x)$. Naaah – that really doesn't sound so good when you say it. I got it! How about $f_{inv}(x)$? That seems to make sense. Well, we'll use it and see if it catches on.

For the purpose of this text, we will define the inverse of the function f as $f_{inv}(x)$, if the following equation holds: $f_{inv}(f(x)) = f(f_{inv}(x)) = x$

Examples of inverse functions

(a) Given $f(x) = 2x - 6$ and $g(x) = \frac{1}{2}x + 3$. Are $f(x)$ and $g(x)$ inverse functions?

$$f(g(x)) = 2g(x) - 6 = 2(\frac{1}{2}x + 3) - 6 = x + 6 - 6 = x$$
$$g(f(x)) = \frac{1}{2} f(x) + 3 = \frac{1}{2} (2x - 6) + 3 = x - 3 + 3 = x$$

So, since $f(g(x)) = x$ and $g(f(x)) = x$, f and g are inverse functions. We can now say that $g(x) = f_{inv}(x)$, or conversely, $f(x) = g_{inv}(x)$. They are both inverses of each other. It's kind of like your siblings – you are your brother's brother, but your brother is your brother too. You are both brothers of each other. The function f is the inverse of g, and the function g is the inverse of f.

Here's something interesting...it's impossible for all siblings in a given family to have exactly one brother and one sister. Go ahead, try it! No matter how you arrange kids, each one cannot have *exactly* one of each.

(b) Given the following table, are $P(x)$ and $Q(x)$ inverses?

x	0	3	5	-2	1
P(x)	1	-2	0	5	3
Q(x)	5	1	-2	3	0

Does $P(Q(x)) = Q(P(x)) = x$ for all x-values in the table?

Let's start with $x = 0$. $P(Q(0)) = 0$ and $Q(P(0)) = 0$...check!

Let x = 3. P(Q(3)) = 3 and Q(P(3)) = 3…check!
    What about x = 5? P(Q(5)) = 5 and Q(P(5)) = 5…check!

You check x = -2 and x = 1 to see if P and Q are inverse functions.

(c) Do the graphs below represent inverse functions? The mirror line is graphed for your convenience.

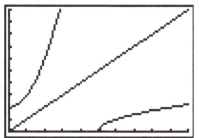

Clearly not. When the upper function is reflected over the line y = x, it does not exactly match the bottom function.

You can also tell by looking at the point (0, 2) on the upper function. When reflected, it would become (2, 0) on the bottom function, however that point is not even *on* the bottom function.

Professor's Practice Pause

Given the following two functions: $f(x) = (x + 1)^2 - 4$ and $g(x) = \sqrt{(x + 4)} - 1$

(a) State the domain *and* range of each function. ((Hint: You may want to graph them))

(b) Find $f(g(5))$ and $g(f(5))$. What do the answers suggest?

(c) Use composition to verify that f and g are inverses of each other

(d) Graph $f(x)$ in Y_1, $g(x)$ in Y_2 and x in Y_3. Copy your TI output below. Do the graphs of these functions seem to verify the same conclusions from (b) and (c)? Why or why not? Ask your professor about this in your next class. In fact, you may want to say something like this, "Professor, would we need to limit the domain of $f(x)$ so that f and g are inverses?"…brownie points!!

Associated End of Booklet exercises are 37 – 52.

Finding the equation of the inverse

It's easy enough to *verify* that 2 functions are inverses using either composition or the TI-83, but if you are given only one function, how do you actually find the equation of its inverse? I bet you asked your roommate the same exact question the other day when you were working on the previous professor's practice pause (Wow!! Now that's *alliteration* for you)

I will show you a simple 4-step procedure for finding the equation of the inverse. It's based on the fact that the inverse is the reflection over the line $y = x$. In other words the x's become y's and the y's become x's. Let's find the inverse of $f(x) = 3x - 7$.

Steps	**Example**
1. Write "y" in place of f(x)	1. $y = 3x - 7$
2. Swap x and y	2. $x = 3y - 7$
3. Solve for y	3. $x + 7 = 3y$…Add 7 to both sides
	$\dfrac{x + 7}{3} = y$…..Divide both sides by 3
	$y = (1/3)x + 7/3$…Break up the fraction
4. Replace "y" with $f_{inv}(x)$	4. $f_{inv}(x) = (1/3)x + 7/3$

To check, we can use composition or start with $f_{inv}(x)$ and apply the same 4 steps to see if we get back to the original function.

Example: Find the inverse of $g(s) = s^2 + 6s + 9$. Let's simply run through the steps…

Step 1: $y = s^2 + 6s + 9$. It's best to factor first…$s^2 + 6s + 9 = (s + 3)(s + 3) = (s + 3)^2$
   $y = (s + 3)^2$ is a better way to write the function
Step 2: $s = (y + 3)^2$
Step 3: $s = (y + 3)^2$
   $\sqrt{s} = y + 3$…take the square root of both sides
   $\sqrt{s} - 3 = y$…subtract 3 from both sides
Step 4: $g_{inv}(s) = \sqrt{s} - 3$

CHECK: $g(g_{inv}(s)) = (g_{inv}(s) + 3)^2 = (\sqrt{s} - 3 + 3)^2 = (\sqrt{s})^2 = s$
   $g_{inv}(g(s)) = \sqrt{g(s)} - 3 = \sqrt{(s + 3)^2} - 3 = s + 3 - 3 = s$
   $g(g_{inv}(s)) = g_{inv}(g(s)) = s$…check!

Topical Summary of Functions

Take about 20 – 30 minutes and create your own summary of this chapter. Go back, review, and write below all of the main points, concepts, equations, relationships, etc. This will help lock the concepts in your brain now, and provide an excellent study guide for any assessments later.

End of Booklet Exercises (the dreaded *EBE*)

1) Define what a <u>function</u> is in mathematics. Use the words *input* and *output* in your definition.

2) What is the Vertical Line Test (VLT)? And how can it be used as a "function detector?"

3) List 3 synonyms for "x-variable" and 3 synonyms for "y-variable."
     x-variable: _____

     y-variable: _____

4) A circle is a common shape that fails the VLT. List 3 other common shapes that would also fail the VLT if graphed on the x-y axes.

5) Sketch several parabolas below. Do they pass the VLT? What does that tell you about all parabolas?

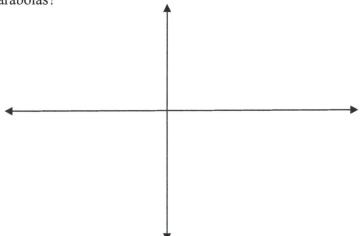

For the statements in questions 6 – 14, circle the input variable and label it with *I*, and circle the output variable and label it with *O*.

6) People's height is a function of people's weight.

7) Your weekly pay depends upon how many hours you work.

8) Your score on an exam is a function of how many hours you study.

9) The area of a circle is a function of the radius.

10) The area of a square is a function of the length of a side.

11) The sales tax you pay depends upon the purchase price.

12) The square of a number is a function of the number.

13) A number is a function of its square.

14) The fractional representation of a number is a function of the its decimal representation.

15) Do the statements in questions 6 – 14 *actually* represent functions? (You may want to consult the text or your answer to question 1). If any are <u>not</u> functions, list them below and explain why not…

16) State whether or not the relationships below are functions. Explain your answer.

(a) $y = \begin{cases} 0, & x \in Q \\ 1, & x \text{ irrational} \end{cases}$

(b) $y = \begin{cases} 0, & x \in Z \\ 1, & x \text{ irrational} \end{cases}$

(c) $y = \begin{cases} 0, & x \in Z \\ 1, & x \in Q \end{cases}$

7) The total cost of a concert (C) is a function of how many tickets you buy (T). If each ticket costs $50, we can write $C = f(T) = 50T$. Which portion of that functional statement is the specific part and which is the symbolic part.

Specific _____   Symbolic _____

18) $A = \{(0,0), (-3, 5), (6, 7), (-10, 6), (x, 2)\}$

    (a) Pick 3 values for x that will make A a function? x = _____

    (b) Pick 3 values for x that will make A *not* a function? x = _____

19) Given $f(x) = -3x^2 + 2x - 7$.
    (a) $f(0) =$ (b) $f(-1) =$

    (c) $f(s) =$ (d) $f(.25) =$

    (e) $f(\pi) =$ (f) $f(\&) =$

    (g) $f(red) =$ (h) $f(1) =$

20) $f(x) = 3(8 - x^2)^{\frac{1}{3}}$ $f(4) =$ _____

21) Write the letter of the corresponding description on the right next to the matching function notation on the left.

    _____ $f(x + a)$

    _____ $f(x) - a$

    _____ $f(x + a) - b$

    _____ $f(x - a)$

    _____ $f(x + a) + b$

    _____ $f(x) + b$

A. Shifts f(x) down a units

B. Shifts f(x) left a units and up b units

C. Shifts f(x) right a units and up b units

D. Shifts f(x) left a units and down b units

E. Shifts f(x) right a units

F. Shifts f(x) up b units

G. Shifts f(x) left a units

22) Given $g(x) = x^3$. Write a new function for $g(x)$ that translates the graph 2 units to the left and 5 units down. Verify your answer by graphing both functions in your TI-83.

23) Given $g(x) = 2x^2 - 1$. Write a new function for $g(x)$ that translates the graph 1 unit to the right and 4 units up. Verify your answer by graphing both functions in your TI-83.

24) The TI-83 screen below shows two functions: 2^x and a translated form of 2^x. What is the equation of the translated function?

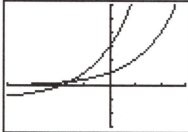

Translated function =

25) Given $f(x) = 3x^2 - 6x + 7$ and $h(x) = x^2 - 4x - 10$

(a) Find $g(x)$ if $g(x) = f(x) - h(x)$

(b) Find $g(x)$ if $g(x) = h(x) - f(x)$

(c) Are the answers to parts (a) and (b) the same? Explain.

26) Using the same functions in question 25)...

(a) Find $g(x)$ if $g(x) = h(x) + f(x)$

(b) Find g(x) if g(x) = f(x) + h(x)

(c) Are the answers to parts (a) and (b) the same? Explain.

27) 3 – 2x + h(x) = 6x + 12. Find h(x)

28) g(s) – s² + 7 = 3g(s) + 1. Find g(s)

29) Given the following functions: $G(\alpha) = -2\alpha^2 + 6\alpha - 3$, $H(\alpha) = 25 + \alpha^3$, $T(\alpha) = \dfrac{\alpha}{3} - 2$, and

$P(\alpha) = \alpha^2 - 8\alpha + 4$. Use either the foil or box methods to find the following products:

(a) G(α)T(α)

(b) P(α)G(α)

(c) $T(\alpha)H(\alpha)$

30) The rectangle below has a length of $x - 3$ and a width of x. A new rectangle is formed by increasing the length by x^2 and the width by $(5 + 2x)$. Express the area of the new rectangle as a function of x. Call your answer $A(x)$. Simplify your answer. (Hint: Draw the new rectangle).

$x - 3$

x

31) Two functions are multiplied using the box method below. There are 3 errors somewhere in the 9 boxes. Circle the errors and write the correct term in the box.

	$3x^2$	$-2x$	4
$-x^2$	$-3x^4$	$2x^2$	$-4x^2$
$-5x$	$-15x^3$	$10x^2$	$20x$
-1	$-4x^2$	$2x$	-4

32) Finish problem 31) with the corrected terms. That is, what is the product of $3x^2 - 2x + 4$ and $-x^2 - 5x - 1$?

33) A study of school size versus student achievement in college algebra courses was conducted to determine if the size of the school made a difference in determining how students faired in algebra.

It turns out that percentage of failures is a function of the student population. The data below represents the size of the student population (in thousands) and the percent of students that failed college algebra their first time.

Student pop (thousands)	0.5	2	4	5.3	6.5	8	9	12	15	17
Percent failures	0.2	2.5	6.2	13	18	29	35	48.5	36	27

(a) What are the input and output variables for this situation?

(b) Does the data truly represent a function? Why or why not?

(c) Write a symbolic equation using "G" as the function name, "s" for the input, and "p" for the output.

(d) Graph the data on your TI-83 using a scatter plot. Sketch your scatter plot below

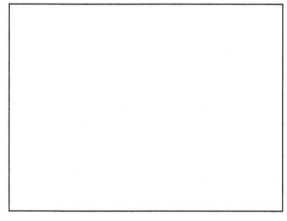

(e) Of the four mathematical models below, which one best matches the data:

    i) $f(x) = 3.2x - 2.4$ *iii)* $f(x) = -.3x^2 + 7x - 11$

    ii) $f(x) = -.057x^3 + 1.2x^2 - 2.5x + 1.7$ *iv)* $f(x) = -10x + 62$

(f) Using your best model from (e) above, what should the percent failures be for a student population of 10,000?

34) Analysts track the price of a gallon of crude oil to be able to spot trends in the price of fuel and make predictions about our economy. The prices of a gallon of crude oil are listed below by year, starting in 1970. Note that 1970 is year 0.

Year	1 gallon of crude oil (US $)
0 (1970)	28
5	46
10	33
15	37
20	45
25	53
30	65
35 (2005)	86

(a) Create a scatter plot with your TI-83 and sketch your plot below fairly neatly. **Label** the axes with the proper <u>variables</u> and <u>numerical scales</u>.

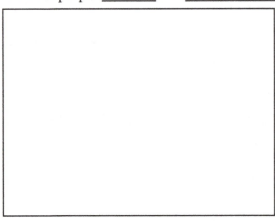

(b) Of the four mathematical models below, which one best matches the data: Circle the letter of the best answer.

i) $f(x) = 1.3x + 25$

iii) $f(x) = 0.07x^2 - 0.61x + 35$

ii) $f(x) = 0.003x^3 - 0.1x^2 + 1.38x + 31.6$

iv) $f(x) = 29 (1.03)^x$

(c) Using your best model from part (b), predict the cost of a gallon of crude oil in the year 2020?

35) The following table shows the average price for 1 gallon of regular unleaded gasoline in the US for various years between 1976 and 2007.

YEAR	PRICE (cents)
1976	61
1980	125
1984	121
1988	95
1992	113
1996	123
2000	151
2004	188
2005	230
2006	259
2007*	275

Source: *www.fueleconomy.gov*
* *estimated*

(a) Create a scatter plot of the data using your TI-83 and sketch it below.

(b). Of the four mathematical models below, which one best models the gasoline data:

i) $f(x) = 40x - 2$

ii) $f(x) = -0.22x^3 + 5.8x^2 - 6.3x + 33.6$

iii) $f(x) = 0.275x^2 - 3.5x + 100$

iv) $f(x) = 100x - 25$

(c). Explain in a few sentences how you found your answer to question (b)?

(d). Using your answer to question (b), what would you estimate as the average price per gallon in 2008? Note that this is an example of <u>extrapolation</u> (projecting beyond the given data).

(e). Use your favorite search engine (Google, Yahoo, Ask, etc) and find the actual average price of a gallon of gas in the US for 2008. How does your estimate compare with the actual value?

(f). What are some reasons for the difference between the two values in question (e)?

36) Consider the situation where the letters of the alphabet are the inputs and the numbers from 1 to 26 are the outputs, as below...

a b c d e f g h i j k l m n o p q r s t u v w x y z
1 2 3 4 5 6 7 8 9 10 11 12 13 14 15 16 17 18 19 20 21 22 23 24 25 26

Now consider a function called Spell(α) that sums all the inputs into a percentage output. For example:

Spell(a) = 1%, Spell(d) = 4%, Spell(y,z) = 25 + 26 = 51%, Spell(a, b, c) = 1 + 2 + 3 = 6%

How about Spell(hardwork) = 8+1+18+4+23+15+18+11 = 98%...and...
Spell(knowledge) = 11+14+15+23+12+5+4+7+5 = 96%...and...
Spell(attitude) = 1+20+20+9+20+21+4+5 = 100%. See that, *attitude* is 100%!

Is Spell(α) really a function? Explain...

37) State the domain and range of the following functions. It may be very helpful to graph them.

(a) f(x) = s^2

$D_f =$ _____

$R_f =$ _____

(b) y = 2x^2 – 4x + 13

$D_y =$ _____

$R_y =$ _____

(c) g(x) = 3x – 5

$D_g =$ _____

$R_g =$ _____

(d) y = x^3

$D_y =$ _____

$R_y =$ _____

(e) h(x) = 1/x

$D_h =$ _____

$R_h =$ _____

(f) z = 1/x^2

$D_z =$ _____

$R_z =$ _____

38) Given $Q = \{(0,0), (-2, 5), (3, 12), (5, 5), (x, y), (\&, *), (-\pi, e)\}$

(a) List the elements of D_Q and R_Q.

(b) Write the set Q_{inv}.

(c) List the elements of D_{Qinv} and R_{Qinv}

(d) What do you notice about your answers in parts (a) and (c)?

39) What is the domain and range of all lines that have non-zero slope?

40) What is the domain and range of all horizontal lines?

41) Do vertical lines have a domain and range? Why or why not?

42) Consider the following 2 functions: $f(x) = x^2 - 6$ and $g(x) = 3 - 2x$

(a) Find $f(g(2))$ (b) Find $g(f(-1))$

(c) Find $f(g(x))$ (do not simplify)   (d) Find $g(f(x))$ (do not simplify)

43) Given $f(x) = x - 1$, $g(x) = 2 - x$, and $h(x) = x^2$. Find $h(g(f(x)))$

44) Given $T(R(B(x))) = x^2 + 1$. Find one possibility for B(x), R(x), and T(x)

45) Given $A(p) = 7 - 4p$ and $T(p) = 2p^3 - 1$, Find A(T(0)).

46) Find the values below using the following two tables:

x	0	1	2	3	4	5
f(x)	-1	3	-2	0	6	10

x	-3	-2	-1	0	1	6
g(x)	17	-5	5	1	2	-3

(a) Find f(3)   (b) Find g(-1)

(c) Find f(g(0))   (d) Find g(f(2))

47) Find the values below using the following two tables:

x	0	2	5	-2	10	-6
F(x)	3	-6	7	0	4	8

x	8	0	10	7	-6	4
G(x)	9	-5	1	0	9	10

        (a) F(G(7)) = _____   (b) G(F(5)) = _____

(c)   F(G(4)) = _____.   Based on this answer and the tables above, would you say that F(x) and G(x) are inverses? Explain why or why not.

48) Given: $A(s) = -s^2 + 6s - 9$, $H(s) = 5s - 2$, $P(S) = \sqrt{s}$ (do not simplify in this question)

(a) Find H(A(s))   (b) Find P(H(s))

(c) Find A(H(P(s)))

49) Find the Volume and Surface Area of the open top, rectangular box below.

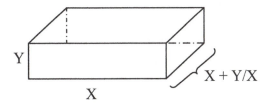

Y

X

X + Y/X

50) Verify that f(x) = 3x − 5 and g(x) = $\dfrac{x}{3} + \dfrac{5}{3}$ are inverse functions using the composition definition.

51) In the middle of a Chemistry lab on "boiling points of liquids", you notice that the measured temperature in Celsius is a function of the temperature in Fahrenheit. You write down the equation as follows: C = T(F) = 5/9 (F − 32)

(a) Identify the input and output variables in this situation.

Input _____

Output _____

(b) What is the mathematical domain and range of the function T?
Note: T(F) is in the form "y = mx + b"

Domain _____

Range _____

(c) Calculate T(212), and explain what that means in this situation.

(d) Find the equation of the inverse of C(F). That is, find $C_{inv}(F)$.

(e) Calculate $T(-40)$ and $C(-40)$. What does this lead you to believe about the two functions? Verify your assumption graphically.

52) For the functions listed below, find the equation of the inverse.

(a) $f(x) = -10x - 3$

(b) $g(x) = 5x^2 + 7$

(c) $h(x) = \sqrt{x - 2}$

(d) $f(x) = x^2 + 6x + 9$

(think about this one a bit)

EXTENDED PROJECTS

1. A look at Manufacturing...

If you have a bottle of Aspirin, Tylenol or Ibuprofen in your room, go check to see how many pills came in that bottle. Now imagine the store where you bought it, and estimate the number of bottles on the shelf at that store...got that number? Ok, now (you knew this was coming) figure out how many of those pills are on the shelf in that store. Now think about how many stores there are in your town, how many towns there are in your state, and how many states there are in the country...there are quite a lot of pills made by that company, huh? Perhaps tens of millions or hundreds of millions each year? According to the website www.madehow.com, Americans consume about 16,000 tons of aspirin each year – which represents almost 30 *billion*, 500mg tablets.

You can imagine that an aspirin manufacturing facility is HUGE!! They probably have big, house-size vats and great furnaces to heat and combine all of the raw ingredients. The process undoubtedly has many steps from raw materials to finished product.

Imagine that you are hired as a new supervisor for one of the production lines that has 5 different stations that raw materials must go through to come out as finished product. Each station has a *transfer function* that lets you figure out the weight of raw material going from one station to the next – as shown below. The variable, W, represents the weight (in pounds) of raw material going into Station 1. $S_1(W)$ is the weight coming out of Station 1, $S_2(W)$ is the weight coming out of Station 2, etc...Notice that the weight at each Station depends upon the previous Station.

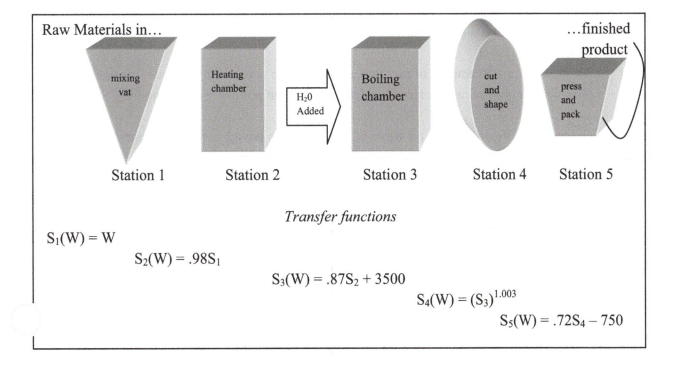

Raw Materials in...   ...finished product

mixing vat   Heating chamber   H₂0 Added   Boiling chamber   cut and shape   press and pack

Station 1   Station 2   Station 3   Station 4   Station 5

Transfer functions

$S_1(W) = W$

$S_2(W) = .98S_1$

$S_3(W) = .87S_2 + 3500$

$S_4(W) = (S_3)^{1.003}$

$S_5(W) = .72S_4 - 750$

(a) In your pre-shift meeting, your manager wants to know the weight of material that comes out of S_5 if the line starts with 15,000 pounds of raw material. Calculate this answer Station by Station until you reach the end.

(b) 15 minutes before your shift begins, the raw material count changes from 15,000 to 22,000 pounds. Your manager wants to know the new output weight. Calculate this answer just like you did in part (a).

(c) After a week on the job, you realize you are doing the same calculations over and over again. A light bulb suddenly goes on over your head!! You remember something from your College Algebra class. Instead of going through all 5 steps each time she wants an answer, you decide to use *composition* of functions to be able to go directly from the input to the output (from the input of S_1 to the output of S_5). Algebraically find $S_5(S_4(S_3(S_2(S_1(W)))))$ and call this new function $T(W)$ – for Total weight function.

(d) Enter $T(W)$ into Y_1 of your TI (remembering to use X, not W), and use the TABLE function in ASK mode to find $T(15,000)$ and $T(22,000)$. Your answers should agree with parts (a) and (b)

(e) What do you notice about the relationship between the pounds of raw material in and the pounds of product out? What are some possible, physical explanations for this?

(f) A new piece of apparatus is added to your line at Station 6 with transfer function given by $S_6(W) = 2(S_5 - 6780)$. Find your new $T(W)$ and use your TI-83 to find $T(15,000)$ and $T(22,000)$ just as you did in part (d). Does this new piece of apparatus at Station 6 tend to increase or decrease the output? Explain.

(g) Use the TABLE function on your calculator to find estimates for the following to the nearest hundred pounds:
     (i) The minimum amount of raw material needed to realistically operate this line.
     (ii) The amount of raw material needed so that *input* pounds = *output* pounds.

(h) Write a 1,000,000 word essay on why you would like to have a career as a pharmaceutical manufacturing supervisor. ((Note: this part of the problem is *optional*))

NOTE: The above process, while not the correct process in making aspirin, does represent a schematic for the production of various, complex products. The production of aspirin is actually easier since it typically contains only 4 ingredients: acetylsalicylic acid, water, corn starch, and a lubricant (such as vegetable oil). Visit the website above and www.aspirin.com to learn about one of the world's oldest wonder-drugs.

2. A look at Wind Chill...(Functions of several variables)

In calculus and other math courses, we study functions of *several* variables, not just one. For example, f(x) might become f(x, y, z). The table below shows the wind chill temperatures in Fahrenheit given the air temperature (T) and the wind speed in miles per hour (V). The equation for Wind Chill is at the bottom of the table. Notice that Wind Chill *depends upon* (is a function of) both T and V (air temperature and wind speed). We can write f = WC(T, V).

NWS Windchill Chart

Wind (mph) \ Temperature (°F)	40	35	30	25	20	15	10	5	0	-5	-10	-15	-20	-25	-30	-35	-40	-45
5	36	31	25	19	13	7	1	-5	-11	-16	-22	-28	-34	-40	-46	-52	-57	-63
10	34	27	21	15	9	3	-4	-10	-16	-22	-28	-35	-41	-47	-53	-59	-66	-72
15	32	25	19	13	6	0	-7	-13	-19	-26	-32	-39	-45	-51	-58	-64	-71	-77
20	30	24	17	11	4	-2	-9	-15	-22	-29	-35	-42	-48	-55	-61	-68	-74	-81
25	29	23	16	9	3	-4	-11	-17	-24	-31	-37	-44	-51	-58	-64	-71	-78	-84
30	28	22	15	8	1	-5	-12	-19	-26	-33	-39	-46	-53	-60	-67	-73	-80	-87
35	28	21	14	7	0	-7	-14	-21	-27	-34	-41	-48	-55	-62	-69	-76	-82	-89
40	27	20	13	6	-1	-8	-15	-22	-29	-36	-43	-50	-57	-64	-71	-78	-84	-91
45	26	19	12	5	-2	-9	-16	-23	-30	-37	-44	-51	-58	-65	-72	-79	-86	-93
50	26	19	12	4	-3	-10	-17	-24	-31	-38	-45	-52	-60	-67	-74	-81	-88	-95
55	25	18	11	4	-3	-11	-18	-25	-32	-39	-46	-54	-61	-68	-75	-82	-89	-97
60	25	17	10	3	-4	-11	-19	-26	-33	-40	-48	-55	-62	-69	-76	-84	-91	-98

Frostbite Times   30 minutes   10 minutes   5 minutes

$$\text{Wind Chill (°F)} = 35.74 + 0.6215T - 35.75(V^{0.16}) + 0.4275T(V^{0.16})$$

Where, T= Air Temperature (°F) V= Wind Speed (mph)   Effective 11/01/01

Source: *National Weather Service*

(a) Let's assume the wind speed is constant at 5 mph. Does WC(T,5) represent a function of air temperature? Why or why not?

(b) Let's assume the air temperature is constant at 10°F. Does WC(10,V) represent a function of wind speed? Why or why not?

(c) Assume both T and V are allowed to vary. Explain why WC(T,V) is a function.

(d) Use the formula to calculate WC(3,27). Does the answer fall where you would expect it to on the chart?

(e) WC(-17,10) = -37. What does this equation mean in terms of the situation?

(f) WC(0, x) = -27. Use the chart to find x.

(g) Let's look at the limitations of the formula. What happens when there is no wind at all (i.e. V = 0)? Use the formula to calculate WC(15,0). Does your answer make sense? Explain what is happening in this situation.

3. A look at Agriculture...

In an apple orchard, there is a definite relationship between the amount of fertilizer applied to trees and the number of bushels of apples produced. See the data below...

(a) Create a scatter plot of your data

Fertilizer (pounds)	Apples (bushels)
0	200
10	300
20	480
40	550
60	420
70	180
80	0

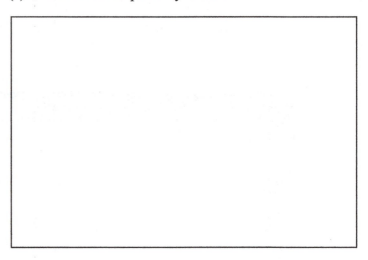

(b) Which equation below best models the data?

    i) $f(x) = 10x + 200$ *iii)* $f(x) = -.01x^3 - .2x^2 + 18x + 180$

    iii) $f(x) = -.3x^2 + 21x + 170$ *iv)* $f(x) = .00001x^4 - .01x^3 + .4x^2 + 10x + 200$

(c) Using the model you choose in question (b), write a specific function statement that means "the number of bushels of apples for 50 pounds of fertilizer." Now find the amount of apples for 50 pounds of fertilizer.

(d) Find the *practical* domain and range of your function from part (b). The practical domain and range are those portions of the mathematical domain and range that make sense in actuality. For example: The domain of a function might be *all real numbers*, however if the input variable is number of seconds, then only positive values of the domain are practical (there's no such thing as negative time!).

(e) For your function, what does the statement f(27) = 518 mean with respect to the variables?

(f) What we have learned above is that bushels (of apples) are a function of pounds of fertilizer. We can write $B = g(F)$ as a symbolic statement. It is also true that fertilizer is a function of cost in dollars…so we can write $F = h(D) = 12D$ (we get 12 lbs of fertilizer per dollar). It would be helpful to find bushels as a function of dollars (eliminating the fertilizer). In other words, find $g(h(D))$.

(g) Answer the following questions based on your answer in (f).

1) $g(h(5)) = 350$. What does this equation mean with respect to the variables?

2) Find $g(h(6.25))$.

3) For what value of x is $g(h(x)) = 0$? What does this mean with respect to the variables?

(h) It turns out that apple production is also a function of labor charges (cost to pick them). Bushels of apples are a function of labor in hours…$B = p(D) = .1D$ (Every \$10 in labor yields 1 bushel). Substitute and simplify the following equation… $B(D) = g(h(D)) + p(D)$.

Booklet on the Linear Family of Functions

Where do Lines live?

Real numbers live on the number line, complex numbers live in the complex plane, termites live in wood…so where do linear functions (lines) live? They live in a place called the x-y plane, or the x-y axes, or x-y coordinate system, or even the *Cartesian* coordinate system. All of these names apply to the same thing – the picture below…

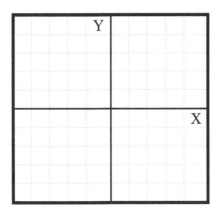

The word Cartesian comes from René Descartes, a 17th Century, French mathematician. The "cartes" in *cartesian* comes from his name. As math legend has it, Descartes, a child prodigy, was looking up at his ceiling one day. While tracking the movements of a spider (or fly), he came up with the foundation for the coordinate system. As Ripley would say, "Believe it…or not?!"

In the x-y plane above, the x-axis is also called the horizontal axis or the input axis (function language), while the y-axis is called the vertical axis or the output axis.

x-axis	y-axis
horizontal axis	vertical axis
input axis	output axis

Just like protons and electrons are the building blocks of atoms, so *points* are the building blocks of lines. Points (or coordinate points) are the little things that populate our x-y axes. This is probably a review for you, but let's define what a point looks like…

$$point = (x, y) = (horizontal, vertical) = (input, output)$$

A point has an x-coordinate and a y-coordinate…just numbers that are plotted on the x-y axes. The middle point, when both x and y = 0 is called the *origin*. The point (0, 0) is the origin. All points are plotted by going to x first, and then y. Below are some examples of plotted points.

Match the following points to their plotted locations on the right:

(0, 0)
(-1, 4)
(2, 3)
(-3, -2)
(2, -1)

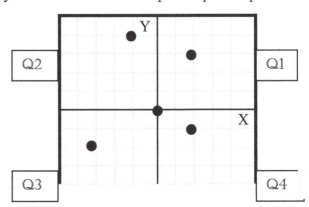

Notice that we have one point in each of the four rectangles that make up the x-y axes. We call those rectangles *quadrants* (like the delta quadrant from Star Trek – where the Borg come from). The upper right hand quadrant is Q1, and then count as you move counter-clockwise from there. Points in the different quadrants have different signs for x and y…

Q1 = (+x, +y)	Q2 = (-x, +y)	Q3 = (-x, -y)	Q4 = (+x, -y)
Examples (3, 6)	(-3, 6)	(-3, -6)	(3, -6)

When you put atoms together, you create molecules; putting musical notes together creates chords; and so it is that putting points together create *graphs*. Sometimes these graphs take on very specific shapes and other times they may be completely random. Below are some examples of *scatter plots* from the TI-83 (plots of variously scattered points).

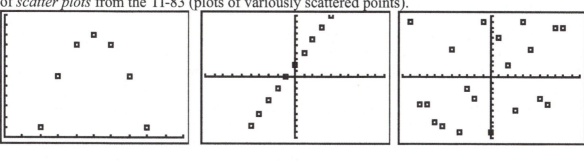

Points in the shape of   Points in the shape of a line   Random points
a parabola

Characteristics of linear functions

Lines are fairly simple…low on the hierarchical ladder of functions…wading at the shallow end of the function pool – you get the idea…easy! There are only 2 characteristics that distinguish one line from another, and those are **slope** and **y-intercept**.

SLOPE:

Have you ever heard anyone say, or maybe have used the expression yourself, "straight line?" If you want to be technical about it, this is actually a redundant phrase. *All* lines are straight. Well, what do we mean by "straight?" That's actually a mathematical term that just means <u>constant</u> slope (or steepness).

Imagine yourself walking up a hill. At the beginning of the hill you lift up your foot to take that first step and your foot meets the pavement at an angle. Halfway up the hill, you take more steps, and your foot is still meeting the pavement at the same angle. Even at the top…same angle. What this means is that your body is experiencing the same (constant) angle, or steepness, or <u>slope </u>throughout the entire journey up the hill.

So the *angle* that the line is at with respect to an axis doesn't matter – all lines are straight since the slope remains constant throughout. All horizontal lines have constant slope, all angled (oblique) lines have constant slope, and even all vertical lines have constant slope. Finding the slope is another matter.

Space for you to draw a curve…

Use the space above to draw a <u>curve</u>; something that goes up and down a little bit (kind of like a little roller coaster). If you have been on a roller coaster, you know the slope is almost never the same. First you're facing up, then down, etc. Now draw a little stick figure representation of yourself on different points of that curve. A <u>curve</u> is not a <u>line</u> simply because it does **not** have constant slope. Your little stick figure man is facing up then down, and at different angles. We sometimes refer to curves as being non – linear.

We use the letter **m** for slope, probably because it comes from the French verb *monter*, which means "to climb."

There are many different phrases for slope…

> The symbol Δ is the upper-case Greek letter "delta," which means "change."

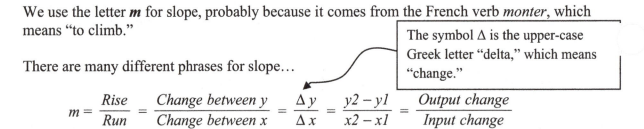

$$m = \frac{Rise}{Run} = \frac{Change\ between\ y}{Change\ between\ x} = \frac{\Delta y}{\Delta x} = \frac{y2 - y1}{x2 - x1} = \frac{Output\ change}{Input\ change}$$

Graphical representation of slope…

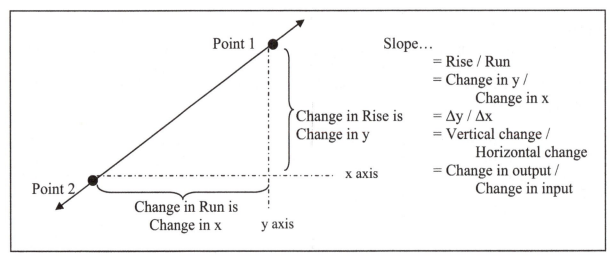

Here's a very important point (VIP)…since the slope of a line is constant, you can find the slope by using <u>any two points</u> on the line. That's right…any two.

Example: To find the slope of the line that goes through the points (1,2) and (5,14), we need to look at how the y values change and then divide that by the change in x values

$$\frac{\text{Change in Y}}{\text{Change in X}} = \frac{\text{(Change from 2 to 14)}}{\text{(Change from 1 to 5)}} = \frac{12}{4} = 3. \text{ So } m = 3.$$

It doesn't matter which point you start at, you will always get the same slope. Think of looking up a hill from the bottom versus looking up the hill from the middle. In both instances you experience the same slope. Try it the other way by starting from (5,14)…

$$\frac{\text{Change in Y}}{\text{Change in X}} = \frac{\text{(Change from 14 to 2)}}{\text{(Change from 5 to 1)}} = \frac{-12}{-4} = +3. \text{ So } m = 3 \text{ again.}$$

Example: Find the slope of the line that passes through (-3, 7) and (2.56, -1.8). If you prefer, you can label the coordinates x_1, x_2, y_1, and y_2, and find the slope that way.

$(x_1, y_1) \quad (x_2, y_2)$

$(-3, 7) \quad (2.56, -1.8) \qquad m = \dfrac{y2 - y1}{x2 - x1} = (-1.8 - 7)/(2.56 - (-3)) = -8.8/5.56 = \underline{-1.583}$

The Distance Formula

Another result that comes from the graphical representation above is the *distance* between any two points. Remember our friend Pythagoras and his theorem? We are really using that to find the distance between any 2 points.

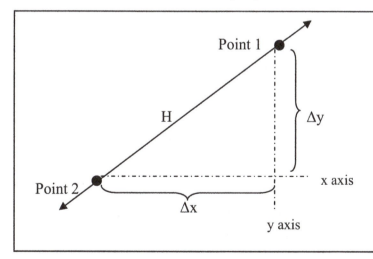

** Notice we have a right triangle. The hypotenuse (H) is really the distance between Points 1 and 2.

The Pythagorean theorem says that the hypotenuse squared is equal to the sum of the squares of the two other sides…so

$$H^2 = (\Delta y)^2 + (\Delta x)^2$$
or
$$H = \sqrt{(\Delta y)^2 + (\Delta x)^2}$$

In other words, the distance between two points is the "square root of the sum of the change in y and the change in x"…

$$D = \sqrt{(y \text{ change})^2 + (x \text{ change})^2} = \sqrt{(\Delta y)^2 + (\Delta x)^2} = \sqrt{(y2 - y1)^2 + (x2 - x1)^2}$$

Example: Find the distance between the points (-2, 6) and (3, 11). We can simply assign x_1, x_2, y_1, and y_2, and substitute into the formula or begin by first finding the two "deltas." Since $\Delta y = 11 - 6 = 5$, then $\Delta y^2 = 25$. Since $\Delta x = 3 - (-2) = 5$, then $\Delta x^2 = 25$ also. The sum of the two deltas is 50, so $D = \sqrt{50} \sim 7.07$

Professor's Practice Pause

1) State which quadrants the following points are in. Draw the points on the x-y plane if you need to.

(a) (-5, -1)   (b) (2, 8)   (c) (π, -e)   (d) (-4, 0)

Q____   Q____   Q____   Q____

2) Fill in the missing x or y value with a number so that the point lies in the quadrant given. There may be several correct answers.

(a) (x, 3) in Q2   (b) (-4, y) in Q3   (c) (-1, y) in Q4

(d) (x, 10) in Q1   (e) (x, y) on the y-axis   (e) (-5, y) on the x-axis

3) Find the slope of the line that passes through each pair of points. Draw the points on the x-y axes if it helps...

(a) (-1, 7) & (3, -2)   (b) (0, 8) & (8, 0)

(c) (-6.25, -9) & (4.5, 10)   (d) (a, b) & (c, d)

4) Find the distance between the pairs of points in (a) – (d) above in question 3.

Associated End of Booklet exercises are 1 – 5

Y-INTERCEPT:

Very simply, the y-intercept is the point at which the line crosses the y (output, vertical) axis. Since this point is always on the y-axis (by definition), the x-coordinate must always be 0. Some of your run-of-the-mill y-intercepts would be the points (0, 4), (0, -2), (0, 10), (0, p), and so on. The y-intercept is a point so it has two coordinates…like all points…(x, y). Because the y-intercept plays an important role in defining a line, we give it a special letter, **b**. Don't ask me why? I haven't been able to figure that one out yet – at least without reasonable uncertainty.

So now we have the slope (m) and the y-intercept (b). Armed with those two bits of information (by the way, two bits is no longer a quarter, but it is a quarter byte)…anyway – armed with those two *pieces* of information, you can conquer any and every question about lines there is in the entire Universe!!!...Hahahaha…(evil laughter)….but more on that in just a little – dare I say it – bit.

The **x-intercept** is the point on the line that crosses the x-axis. So, this time, the y-coordinate must be 0. Examples of x-intercepts are (3, 0), (-4, 0), (1, 0), and (-1207.873, 0). The *origin* (0, 0) is the point that is both the x- and y-intercept.

Example: The table below shows the speed of a test car as it tests a new braking system on a closed track. At time 0, the driver slams on the vehicles brakes and comes to a complete stop 4 seconds later. Do not try this in your living room! The TI scatter plot of the data is shown to the right with the line connecting all 5 points.

seconds	car speed (mph)
0	60
1	45
2	30
3	15
4	0

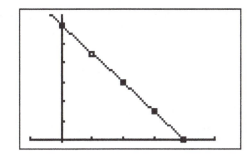

The y-intercept of the line is the point (0, 60). At 0 seconds, the moment the stopwatch is turned on, the vehicle is going 60 mph.

The x-intercept of the line is the point (4, 0). At 4 seconds, the car has come to a complete stop.

Finding intercepts from an equation…very simple. To find the y-intercept, find f(0) – that is, make x = 0 and solve for y. To find the x-intercept (also called the *root* of the equation), make y = 0 and solve for x.

Keep in mind that y *is* f(x) They are interchangeable.

Example: Find both intercepts of the function 3f(x) = 2x – 10.

(a) To find the y-intercept, set x = 0. 3f(x) = 2(0) – 10 = -10. Now divide both sides by 3 and we get f(x) = -10/3. The y-intercept (b) is the point (0, -10/3).

(b) To find the x-intercept, set y = 0. 3(0) = 2x – 10. Adding 10 to both sides gives 10 = 2x, so x = 5. The x-intercept is the point (5, 0)

By putting the slope and y-intercept together in an equation, we can write the general equation for all linear functions as

$$f(x) = mx + b$$

Memory break…what is a coefficient?? The number in front of the variable.

The slope (m) is always the *coefficient* of the input variable (in this case, x) and the y-intercept (b) is always the constant term.

Since the y-intercept always occurs when x = 0, we have f(0) = m(0) + b = b. f(0) = b. In function language, we can write the coordinates of the y-intercept as $(0, f(0))$.

Example: Given the linear functions (a) f(x) = 3x – 5, (b) g(p) = -7p, and (c) H(Z) = (½)Z + 10, and (d) f(x) = 8 + 2x, identify the slope and y-intercept of each function.

(a) In f(x) = 3x – 5, the slope is 3 and the y-intercept is (0, -5)

g(p) could be written as: g(p) = -7p + 0. This way shows that b is clearly 0.

(b) In g(p) = -7p, the slope is -7 and the y-intercept is (0, 0)

(c) In H(Z) = (½)Z + 10, the slope is ½ and the y-intercept is (0, 10)

(d) In f(x) = 8 + 2x, the slope is 2 and the y-intercept is (0, 8) even though the terms are written in a different order. Slope is <u>always</u> the coefficient of x. Y-intercept is <u>always</u> the constant.

Practical meaning of slope and intercept

In real life situations, the slope of a linear function can be considered the "something per something." For example: price per pound, miles per hour, students per class, feet per second, etc.

The y-intercept is the initial value. For example: one-time costs, fixed fees, entrance charges, etc.

Example : Think of how most people pay a university to take classes. Let's say you pay $1000 for every credit hour you take, and you also pay $875 for all the one-time charges associated with that year (student fees, insurance, lab fees, textbooks, etc.).

A function that describes your total cost, T, based on the number of credits you take, c, would be written as T(c) = 1000c + 875.
The variable is c because the number of credits can vary.
The slope is 1000 (the price per credit).
The y-intercept is T(0) = 875 (the fixed costs)

Example: If you attend a juice bar club and the cost to get in is $10, and the price per juice drink is $2.50, you can write an equation for the amount of money you will spend (M) based on the number of drinks you buy (d): M(d) = 2.5d + 10.

The variable is d because the number of drinks can vary.
The slope is 2.5 (the price per drink).
The y-intercept is M(0) = 10 (the one-time entrance fee)

Different types of linear functions

There are 4 special cases of lines that I'll mention here as well: Vertical, Horizontal, Parallel, and Perpendicular.

Place your left arm from elbow to hand horizontally in front of you (m=0). Now slowly begin to elevate the hand leaving the elbow fixed. That's it – you got it. When you get to about 45 degrees, that is a slope of 1 (m = 1). Now keep raising it slowly. What is happening to the value of m? (It keeps getting bigger and bigger doesn't it?)

When your arm is vertical, the slope is at its maximum value (skiers call this a cliff!). What is the highest math number you know? Infinity. So a vertical slope actually means that the slope is <u>infinite</u>. Since all points on a vertical line have the same x-coordinate, the equation of that line is simply *x = x-intercept*. Example, the vertical line x = 3 is shown below:

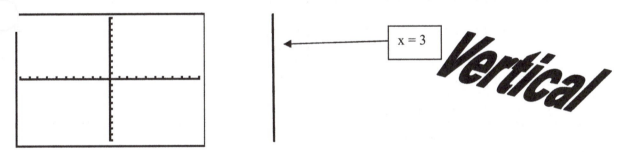

Ok, place your left arm in front of you again. Now bring it down so it is horizontal (skiers call this cross-country…very boring!) This slope is the smallest value you know, which is 0. A horizontal line has 0 slope and the equation is simply *y = y-intercept*. Example, the horizontal line y = 3 is shown below:

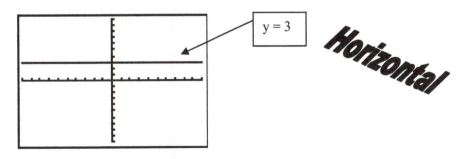

If two lines are drawn that *never* intersect, they are called parallel lines. For example, the floor in your math classroom is parallel to the ceiling there (if it is a flat ceiling). They will never meet if extended. Columns on a Victorian porch are parallel. If extended, they will never cross. In order for 2 lines to be parallel, they must have **_exactly_** the same slope.

The following two lines are parallel: $f(x) = 1.5x - 8$ and $f(x) = 1.5x + 10$. In fact, all lines that have m = 1.5 are parallel, regardless of the value of the y-intercept. The TI plot below shows several lines, all with slope 1.5.

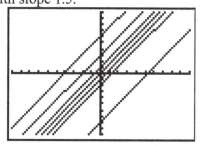

Remember our phrase, "family of functions?" Since all equations of the form $f(x) = mx + b$ form a family, we could say that all equations of the form $f(x) = 1.5x + b$ form a sub-family or nuclear family. (Not *nucular*! I love President Bush, but the word is *nuclear*).

None of these lines above will ever intersect using the axioms of Euclidean geometry. In Non-Euclidean geometry, however, things can change. (Ask your Professor what that means)

The following two lines are *not* parallel: $f(x) = 2x - 7$ and $f(x) = 2.001x - 6$. Even though the slopes are very, very, oh-so close, these lines will still intersect at some point. These two lines happen to intersect at the point (-13000, -26007). We'll be able to figure that out in the Booklet on Several Linear Families.

Math ↔ English

The symbol for parallel lines in math is ‖ like the two l's in para*ll*el.

The symbol for perpendicular in math is ⊥ like the floor meeting the wall.

Speaking of perpendicular, lines are ⊥ if they intersect at 90°. In order for them to meet at right angles, their slopes must be *negative reciprocals* of one another. In other words, "change the sign and flip it over." If a certain line has m = 4, then every line ⊥ to that line will have m = -¼. If another line has slope m = -(4/5), then every line perpendicular to that line will have slope m = 5/4…change the sign, flip over the number…negative reciprocal. Below are the perpendicular lines y = 2x and y = -0.5x +5. We used the Zoom Square function by pressing **ZOOM** and then **5:**. Using the normal rectangular window will cause the lines to look *un*-perpendicular.

Another way to tell if two lines are perpendicular is if the product of their slopes = -1. If the slope of one line is *m*, then the slope of a ⊥ line must be *-1/m*.

$$m*-1/m = -m/m = -1$$

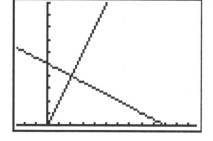

Group Pause

Work in groups of 2 or 3 to solve the following practice exercises. Help each other out, hum a song while you work, and have fun!

1) Find the x and y intercepts of the given linear functions.

(a) $f(x) = x - 7$

(b) $5y - 4 = 3x$

(c) $-3x + 1 = 2f(x)$

(d) $y = -x$

2) Put the functions in question 1) into $y = mx + b$ form and state the slope of each line. You may have already done some of that work...

(a)

(b)

(c)

(d)

3) In the Maltovian province of Irkutsk, drivers must pay to even enter a gas station. The entrance cost is 27 Russian Roubles (RR) and the diesel is 15.75 RR per liter. Write a function statement for the cost of diesel as a function of the number of gallons, and identify m and b for your function.

4) Given the linear function, $P(s) = -2s + 4$.

(a) Write the equation of 2 different functions || to the given function.

(b) Write the equation of 2 different functions ⊥ to the given function.

(c) A horizontal line passes through the point (0, -3). What is the equation of this line?

(d) A vertical line passes through the point (6, 0). What is the equation of this line?

Associated End of Booklet exercises are 6 – 13

$f(x) = mx + b$ does it all!

There are many different forms for the equation of lines. They are as follows:

the *General* form (Ax + By + C = 0),
the *Standard* form (Ax + By = C),
the *Slope/Intercept* form (y = mx + b),
    the *Point/Slope* form (y − y₁ = m (x − x₁)),
    the *Intercept* form (x/c + y/b = 1),
      the *Two-Point* form (I won't even list that equation), and
      the *Parametric* form (even worse....)

> Notice that the *Slope/Intercept* form of the line is the **only** form in which both the slope and y-intercept are given, and these are the 2 most important characteristics of a line.

As you can see, much ado has been made about lines. I will agree that some forms are more efficient for certain calculations, but the major drawback is that students (that includes you) get confused when bombarded with even 2 or 3 of these forms. So, I only teach one form...Hmmm...can you guess which one that is? Yes...the Slope/Intercept form....y = mx + b, or f(x) = mx + b.

Any calculation or manipulation that you will ever have to do with lines can be done with this form. I'll prove it to you! For the next few pages we will look at different styles of questions that I'm sure you have all done somewhere in your mathematical history, and we will answer all of them with y = mx + b.

Example: Find the equation of the line that passes through (1, 2) and (5, 14). We have conveniently drawn these two points below using the scatter plot function of the TI-83...

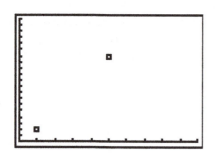

Very Important Point #1 (VIP 1)...

Every line is defined by 2 and only 2 points. When drawing points, thou shalt count to two, no more, no less. Two shall be the number thou shalt count, and the number of the counting shall be two. Three shalt thou not count, nor shalt thou count to one, excepting that thou then proceed to two. Four is right out!

    *borrowed from Monty Python

If you find a way of connecting two points with more than 1 different line, you should consult your math professor and bestow upon him the honor of co-authoring a paper with you!

To find the equation of a line given two points, always take these steps:
    (a) find the slope
    (b) plug in one of the points on the line for x and y
    (c) solve for the intercept

(a) m = Δy / Δx = (14 − 2) / (5 − 1) = 12/4 = 3. Now that we know the slope we can write

$$f(x) = 3x + b$$

> But what is the b?....read on...

Very Important Point #2 (VIP 2)...

"Any point that lives on the line (or curve) must satisfy the equation of the line (or curve)."

This means that a point that lies on the line makes the equation of the line true (satisfies means to make true). So to find "b," we just have to plug in one of the two points that we know are on the line. Let's arbitrarily choose (1, 2)...

$f(x) = 3x + b.$ Substitute 1 for x and 2 for y
$2 = 3(1) + b$
$2 = 3 + b$ subtract 3 from both sides
$b = -1$

> Given a logic statement (if P then Q), the Inverse of that statement is (if *not* P then *not* Q). The Inverse of a true statement may or may not be true. In this case, it is true.

The equation of the linear function that passes through (1, 2) and (5, 14) is **f(x) = 3x – 1**.

The inverse of **VIP 2** also holds true. "If a point does *not* live on the line (or curve), it does *not* satisfy the equation – it makes the equation false."

The point (7, 8) does not lie on y = 3x – 1. How do I know? Plug it in:

Does $8 = 3(7) – 1$?
Does $8 = 21 – 1$?
Does $8 = 20$?   No, so (7,8) is not on the line.

Let's verify our answer by putting into **Y₁**, pressing **GRAPH**, and see if it connects the two points...

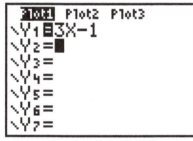
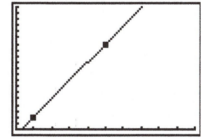

Variations on a theme...

Example: Find the equation of the line that passes through (-1, 5) and is parallel to the line y = 4x – 6.

We use the same three steps, except step (a) is much easier. Since the line we are finding is parallel to the one given, we know they have the same slope. Right away, m = 4.

$f(x) = 4x + b$. Now we pick the one point that we know is on the line (-1, 5) and plug it in.
$5 = 4(-1) + b \rightarrow 5 = -4 + b \rightarrow b = 9$. Our function is f(x) = 4x + 9. We verify on the TI by graphing the given point, the given line, and the equation we found.

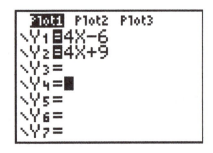
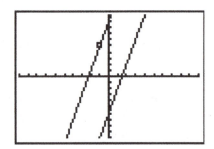

And finally…

Example: Find the equation of the line that goes through (0, 8) and is perpendicular to f(x) = -.5x + 3

Again we know the slope from the given information. The negative reciprocal of -.5 is +1/.5, which is 2. m = 2. So far we know f(x) = 2x + b

We could do the next two steps, but you notice something about the point…yes…the point is the y-intercept already since the x-coordinate is 0. b = 8

Our answer…**f(x) = 2x + 8**. Let's verify…

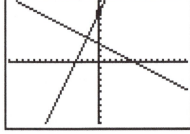

Changing to y = mx + b form

If you happen to presented with a line in another form, no need to panic! You can always change to y = mx + b form by solving for y.

Example: Let's say we have 3x – 4y = 10. This line is in Standard Form. The slope is *not* 3 and the y-intercept is *not* 10, because the line is not in *y = mx + b* form. So, we solve for y.

$$3x - 4y = 10$$
$$\underline{-3x \qquad -3x} \quad \text{…..subtract 3x from both sides}$$

$$\frac{-4y}{-4} = \frac{-3x + 10}{-4} \quad \text{…..divide both sides by -4}$$

$$y = (-3/-4)x + (10/-4) = \textbf{(3/4)x} - \textbf{10/4} \text{ (or .75x} - 2.5)$$

So, the slope of the line is 3/4 and the y-intercept is the point (0, -10/4)

Example: A study done by Virginia Tech. showed the effect of the number of alcoholic drinks on a person's Blood Alcohol Content (BAC). The weight under study was 90 lb. The study also noted that a .01% reduction in BAC was realized for every 40 minutes that elapsed. The data is linear.

Number of Drinks	BAC (%)
1	.05
2	.1
3	.15.
4	.2
5	.25

I am assuming that number of drinks is the input variable and that BAC is the output variable. Does that make sense?

Let me show you several things: (a) find the linear function that represents the data, (b) identify the slope and intercept from the equation, (c) explain the significance of *m* and *b* in relationship to the scenario, (d) find the domain and range of the function, (e) Use the formula to predict a 90 lb. person's BAC if they consume 7 drinks in 2 hours. Wow, that should keep me busy…

(a) It's given that the data *are* linear (technically the word "data" is plural. "Datum" is singular, so we should really write "data are" and "datum is," but no one ever does. The phrase "data is" has been adopted into our society simply because of abundant usage). We pick two points and go through our 3 steps from above…let's pick the first and last points (1, .05) & (5, .25)

$$m = \text{(change in y)/(change in x)} = (.25 - .05) / (5 - 1) = .2/4 = .05$$
So far we have f(x) = .05x + b. Now pick (1, .05) and plug in for x and f(x)
.05 = .05(1) + b → .05 = .05 + b → b must be 0.
Our function is **f(x) = .05x**

(b) The slope is .05 and the y-intercept is the point (0, 0).

(c) A slope of .05 means that for every one alcoholic beverage, a 90 lb. person's BAC will increase by .05%. The significance of the y-intercept is that consuming no alcohol means no alcohol is in your blood – that make sense! 0 drinks, 0 BAC.

(d) The *domain* of every linear function is all real numbers (remember a vertical line is not a function, right?). The *range* of any linear function with non-zero slope is also all real numbers (the range of a horizontal line is simply the y-intercept). Even though the domain and range are both all real numbers, that doesn't help us with the given situation. For example, x = 1000 drinks is in the domain, and the corresponding range value would be .05(1000) = 50% alcohol -- but that's an impossibility in real life. We have something called a **practical domain** and **practical range**. These sets are 'what makes sense' for the given variables. They are also subjective at times. I might say that a practical domain for this situation is $0 \le x \le 4$, which means that a corresponding practical range would be $0 \le y \le .2$ because a 90 lb. person with a .2 BAC is going to be out cold!

(e) Find f(7). f(7) = .05(7) = .35. However we also know that for every 40 minutes, the BAC goes down by .01%. In 2 hours, that would be a .03% reduction, or .35 – .03 = .32 BAC. This person is in serious trouble!!

** Finally, being a parent, I ask that you always consume alcohol responsibly and in moderation. There are many different factors that affect BAC, including weight, gender, amount of food in your body, your emotional state, and other medications you may be taking. Please never drink and drive! And stop texting and driving, too!

Professor's Practice Pause

1) Find the equation of the line that passes through the points (-2, 9) & (1, -7). Use the 3 steps.

2) Find the equation of the line that is parallel to $f(x) = -5x - 7$, and passes through (4, 6)

3) Find the equation of the line that is perpendicular to $f(x) = -5x - 7$, and passes through (4, 6)

4) Change the following line into $y = mx + b$ form and state the slope and y-intercept:
$$-5x + 2y = 12$$

5) Your accountant tells you that the number of IRS audits each year is a function of the number of forms that individuals file. She also tells you that this relationship is linear. What is the input variable and output variable? What would be a reasonable *practical* domain and range for this function?

Associated End of Booklet exercises are 14 – 21

Solving more difficult linear equations

Now that you've been practicing your linear equations, we can begin to stretch you a little with some more difficult linear equations, bringing to bare all of your PEMDAS and fraction skills also.

The key to math is this...Try something! If it doesn't work, try something else!!

Example: Solve $3(x - 2) = -7(x - 1)$. Remember your math grammar...one step at a time, one change per step, and comment each step

$$3(x - 2) = -7(x - 1)$$
$$3x - 6 = -7x + 7 \quad \text{........Distribute to remove parentheses}$$
$$\underline{+7x \qquad +7x} \quad \text{........Add 7x to both sides}$$
$$10x - 6 = 7$$
$$\underline{\qquad +6 \quad +6} \quad \text{........Add 6 to both sides}$$
$$\underline{10x} = \underline{13}$$
$$10 \qquad 10 \quad \text{......Divide both sides by 10 to isolate the x}$$

$$x = \mathbf{13/10} = \mathbf{1.3} \qquad \text{........Solution}$$

Check! $3(1.3 - 2) = -7(1.3 - 1) \ ...\rightarrow... \ 3(-.7) = -7(.3) \ ...\rightarrow... \ -2.1 = -2.1 \ \sqrt{}$

Example: Solve $\dfrac{x + 3}{4} = \dfrac{x - 5}{3}$. The strategy here is to get a common denominator and then eliminate the denominator completely.

$$\frac{3}{3} \cdot \frac{x + 3}{4} = \frac{x - 5}{3} \cdot \frac{4}{4} \quad \text{....Since the common denominator is 12, multiply by (3/3) on}$$

the left side and by (4/4) on the right side

$$\frac{3(x + 3)}{12} = \frac{4(x - 5)}{12} \qquad \text{Since the denominators are equal, we can eliminate them!}$$

$$3(x + 3) = 4(x - 5) \ ...\rightarrow... \ 3x + 9 = 4x - 20 \quad \text{....Distribute to eliminate parentheses}$$
$$\underline{-3x \qquad -3x} \quad \text{.....Subtract 3x from both sides}$$
$$9 = x - 20$$
$$\underline{+20 \qquad +20} \quad \text{....Add 20 to both sides}$$
$$\mathbf{29 = x} \quad \text{........Solution}$$

Check! $(29+3)/4 = (29 - 5)/3 \ ...\rightarrow... \ 32/4 = 24/3 \ ...\rightarrow... \ 8 = 8 \ \sqrt{}$

Example: Solve 2.56T – 8.39 = -.02T + 7.84.   Let's move the T's to the left side of the equation and the numbers to the right side.

$$2.56T - 8.39 = -.02T + 7.84$$

+.02T	+.02T	…..Add .02T to both sides
2.58T – 8.39 = 7.84		
+ 8.39 +8.39		….Add 8.39 to both sides
2.58T = 16.23		
2.58 2.58		……..Divide both sides by 2.58 to isolate T

$$T = 16.23/2.58 = 6.29 \text{ (rounded to 2 decimal places)}$$

Check!   2.56(6.29) – 8.39 = -.02(6.29) + 7.84
            16.102 – 8.39 = -.126 + 7.84
               7.712 = 7.714   √ Or is it? Why are they slightly different??

Example: Solve $\dfrac{3(z-2)}{5} + \dfrac{z}{6} = \dfrac{z}{6} + \dfrac{3z-6}{5}$ Let's look before we leap!

Since 3(z – 2) = 3z – 6, the left and right hand sides have exactly the same stuff, which means they are equal. If we subtracted everything from both sides, we would get 0 = 0. This tells us that we have an *identity* and all values for z will be a solution. The simplified form of an identity is x = x. Any value of x makes the equation true. The word identity comes from identical – both sides exactly the same.

Example: Solve 2x – 4 = 2(x + 3)….Again, distribute and CLT
                    2x – 4 = 2x + 6 ….Distributed the right hand side
                   -2x   -2x   ……Subtract 2x from both sides

            -4 = 6..?? This can't be! If the variables have been eliminated and only ridiculous numerical equations are left, then there are *no solutions*. We could have seen this right away by viewing both sides of the equations as lines…

            2x – 4 has slope = 2   and   2x + 6 also has slope = 2

The left side is parallel to the right side and never the twain shall meet! As parallel lines do not intersect, the equation has no solution.

GROUP PAUSE

CLUE: "A math symbol that is counter-cultural"
ANSWER: ___ ___ ___ ___ ___ ___ ___

Directions: The answer to each equation below is a positive integer. By associating each answer with a letter of the alphabet (A = 1, B = 2, ... , Z = 26), you will create seven letters. Unscramble those seven letters to answer the clue.
Be sure to CHECK your answers by substituting back in the original equations!!

1. $3x + 1 = 4$

2. $c + \dfrac{21 + (c - 9)}{4} = 6 + c$

3. $\dfrac{t - 5}{4} - \dfrac{t}{2} = \dfrac{18 - 5}{4} - \dfrac{t}{2}$

4. $x + 1.24 - 0.07x = 9.61$

5. $4 - (6x + 6) = -(-2x + 10)$

6. $\dfrac{a - 5}{2} = \dfrac{3a}{4} + \dfrac{a - 25}{6}$

7. $5x + 3 = 2(x + 6)$

Modeling linear functions on the TI-83

In the Booklet on Functions we took a look at the Data entering mode and the scatter plot mode of the TI-83. These steps and TI screens are listed here below with some example data for your convenience…

A. Data Entering Mode

1. Press the **STAT** button
2. Press **1** or **ENTER** (to get the EDIT mode)
3. Enter your x (input) data into a List, and your y (output) data into another list
    (be sure there are the same number of data in each list)

B. Scatter Plot Mode

4. Press the **2nd** and **Y=** (to get the STATPLOT mode)
5. Press **1** or **ENTER** (to get into PLOT 1)
6. Highlight the following on the Menu…
    ON
    TYPE (select the first one…scatter plot)
    XLIST: L1, or whatever list you put the x data in
    YLIST: L2, or whatever list you put the y data in
    MARK: Pick one (it doesn't matter)
7. Press the **WINDOW** button…make sure your window _covers_ the data
8. Press **Y=**…and clear unwanted equations
9. Press GRAPH.

> The Data Trick…
> If you're x-data is in years, instead of starting at year 2000 (for example), start at year 0 when you enter the data into the TI-83 list. Use 1 for 2001, and 2 for 2002, etc.

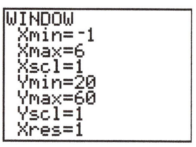

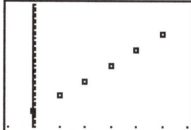

We now want to add a third step called the _Regression Equation Mode_. The word _regression_ might seem familiar to you from your psychology class, remember? _Close your eyes. I'm going to count to 5. You're starting to feel more and more relaxed._ Hypnosis! The Sigmund Frued character would _regress_ you back to your childhood using hypnosis to try and unravel your psyche…

Unfortunately, the math use of the word regression means absolutely nothing like the psychology use for the word.

Fundamentals of regression

Let's back up a bit first. Many of the equations and math ideas you will be studying get messed up by "real life." Real data is not nice and predictable and there are always more assumptions and considerations than simply plugging numbers into a formula will allow. That is why mathematicians create <u>mathematical models</u> using regression equations in fields such as economics, engineering, physics, meteorology and marketing (to name just a few). These models help show how things are behaving over time, and they also allow us to predict future values (extrapolation).

Now let's break down the phrase "linear regression." Linear means line and regression means using a math model. So, linear regression means finding a linear equation that we can use to model a set of data or a relationship between two variables. We also call this line the <u>trend line</u> or the <u>line of best fit</u>, because it is the equation that produces the best results and minimizes the amount of error between the actual data and the equation itself. This means if we were to plot our data (the two variables) on the coordinate axes, our line of best fit would be closest to the most points than any other line.

Just how do we *measure* best fit? When we find the regression equation on the TI-83, the calculator not only gives an equation, but an "**r**" value as well. This **r** – value is called the <u>correlation coefficient</u>, and it is a measure of just how good the model fits the data. The values of **r** can range from -1 to $+1$. If the absolute value of **r** is close to 1 the model will be an accurate predictor. As the value of **r** gets closer to 0, the model (and the prediction) become less accurate.

If **r** is positive it means the relationship between our two variables is direct, the line itself has positive slope, and we say the two variables are *positively correlated*. In other words, if one variable goes up the other goes up, and if one variable goes down the other goes down. An example of a <u>positive correlation</u> is the relationship between numbers of alcoholic beverages consumed versus blood alcohol content (BAC). The more someone drinks, the higher that person's BAC will be. Another example (based on educational research) is the amount of education versus an individual's starting salary. It has been shown that people with higher levels of education will begin their careers at a higher salary.

A negative **r** means there is an indirect relationship – as one variable goes up the other goes down, and vice-versa. The slope of this line is negative and we say that the two variables are *negatively correlated*. Two variables that would exhibit a <u>negative correlation</u> are amount of cigarettes smoked versus lung capacity, or applied antiseptic versus microbial growth. In each case, when the input variable (cigarettes smoked and applied antiseptic) increases, the measured or output variable (lung capacity and microbial growth) decreases.

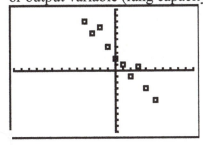 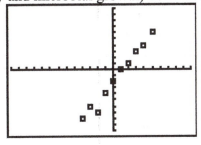 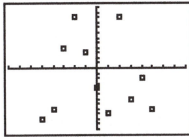

Linear data with r < 0 **Linear data with r > 0** **Random data with r ~ 0**
(negative correlation) (positive correlation) (no correlation)

There are some disadvantages or limitations to using linear regression. This model is very susceptible to <u>outliers</u>. Outliers are those points that don't follow the pattern described by the rest of the data points – they "lie out" of the general trend. As you will see in the Booklet on Logarithmic Functions, simply moving the position of an outlier not only changes the r-value of a model, but can changes the family of functions used to model the data as well. Because of this, some researchers and data collectors simply remove them altogether.

I present you with the Linear Regression Equations…y = mx + b, where…

$$m = \frac{n \Sigma x y - \Sigma x \Sigma y}{n \Sigma x^2 - (\Sigma x)^2} \qquad \text{and} \qquad b = \frac{\Sigma y \Sigma x^2 - \Sigma x \Sigma x y}{n \Sigma x^2 - (\Sigma x)^2}$$

Now you can see why we don't calculate these by hand!! So, we turn to our trusty, stupid, but very fast friend – the TI-83. Continuing from the Data Entering mode and the Scatter Plot mode…

C. Regression Equation Mode

10. Press the **STAT** button
> 11. Right Arrow over to CALC
> 12. Press **4:** (for Linear Regression)…Note that there are other models here also
>> At this point you should see LINREG (ax + b) on your Home Screen
> 13. Type your x and y lists in …
>> For example, if you used L1 and L2, type L1, L2….that's **2nd 1 COMMA 2nd 2**
> 14. Press **ENTER**

You should see the following….
> y= ax + b
> a = number
> b = number
> r^2 = number
> r = number

> **NOTE**: If you do not see r or r^2, do the following….
>
> Press **2nd 0** (which is the CATALOG mode)
> Arrow down until you see DIAGNOSTIC ON
> Press **ENTER** and then **ENTER** (it should say DONE)
> Go back and start at step 10

Let's tie this all together by walking through an example…..

The "troop surge" is a phrase commonly used to describe President George W. Bush's plan to increase the number of American troops deployed to the Iraq War to provide security to Baghdad and Al Anbar Province. The table below shows the US death toll per month beginning two months after the official start of the surge. May 2007 is month 0, June 2007 is month 1, etc.

Month	Total US fatalities *
0	126
1	100
2	73
3	80
4	65
5	40
6	37
7	23

* These include both combat related and non-combat related deaths.

Source: www.usatoday.com/news/world/iraq/2007-05-29-troop-deaths_N.htm

We'll start by entering the data into the TI-83. Place the month data in L_1 and the deaths in L_2.

```
L1          L2          L3       3
0.000       126.00      ▆▆▆▆▆
1.000       100.00
2.000       73.000
3.000       80.000
4.000       65.000
5.000       40.000
6.000       37.000
L3(1)=
```

Now to the scatter plot... *((I should note that it is not always necessary to create a scatter plot. If you just need the equation you can enter the data and skip right to the last mode))*

Turn on a STATPLOT by pressing **2nd Y=** and step through the menu. Adjust the window to cover the data.

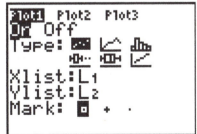

Display the scatter plot by pressing **GRAPH**...

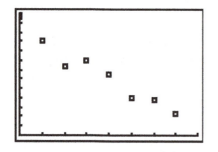

Just by inspection we know the following:

(a) The data appear to be linear
(b) The variables are negatively correlated
(c) The r-value and the slope are both negative.

Now we can find the equation by pressing **STAT**, right arrow over to **CALC** and press **4:**. You should see *LinReg(ax + b)* **and** the blinking cursor. Add the lists... **L₁ COMMA L₂**. Press **ENTER.**

EDIT **CALC** TESTS	LinReg(ax+b) L₁,	LinReg
1:1-Var Stats	L₂	y=ax+b
2:2-Var Stats		a=-13.690
3:Med-Med		b=115.917
4:LinReg(ax+b)		r²=.944
5:QuadReg		r=-.972
6:CubicReg		
7↓QuartReg		

Again, if you do not see the r^2 or r values, go up a few pages to the note I wrote you before and complete those steps. You will have to "run" the regression over again from the STAT option.

It would be nice to see how the equation of the line "fits" or "matches" the data. The r-value of -.972 tells us that we have a very good match...remember the highest value in this case would be -1. You can think of the r-value as *percentage of match*, although that is not an accurate description mathematically...but it helps! In our example, we can say that we have a 97% match between data and model. Your professor may not like this notion, in which case we discourage you from applying it ☺

A way to have the TI place the equation into Y1 would be nice to know, especially when we hit the Booklets on polynomials, where the equations will be like, totally out of control!

Press **Y=** and make sure the cursor is in Y_1. All of the other places should be cleared. Press the VARS button and then press or arrow down to **5: Statistics**. Right arrow over to **EQ** and press **1:**. You should see the equation in Y_1. Now press **GRAPH**. Violá!

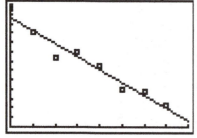

Because the model matches the data quite nicely – visually and according to the r-value, we should feel comfortable using our equation to *predict*…remember that was the whole **goal** of mathematical modeling (I smell quiz question!!)

Question: Using our model, what should the fatality level be in January 2008, the 8th month after the start of the surge (i.e. x = 8). We can easily set up a table to figure that out…

```
 X       Y₁
 2.000   88.537
 3.000   74.847
 4.000   61.157
 5.000   47.467
 6.000   33.777
 7.000   20.087
 8.0000  6.397

X=8
```

Our model predicts about 6 or 7 deaths in January 2008. We actually know how many deaths occurred for that month…there were approximately **35**! Huh?? What does that tell us?

    (1) First of all, we must have a reasonably good match (high r-value) in order to even *begin* making predictions.

    (2) Even if the r-value is good, the future is the future, and it is sometimes unpredictable.

    (3) Interpolations are typically more accurate than extrapolations. If we were to predict a mid-month value in the above data, like 4.5, our interpolation would be very accurate.

    (4) The further you extrapolate, the less accuracy you will have in your predictions. Adjustments to your model may be needed.

Professor's Practice Pause

The data below represents the Dow Jones Industrial Average closing figure for business days in March, 2008. Letting March 3rd be day 0, March 4th be day 1, and so on, create a scatter plot of this data and sketch it in the space to the right.

DATE	CLOSE
31-Mar-08	12,262.89
28-Mar-08	12,216.40
27-Mar-08	12,302.46
26-Mar-08	12,422.86
25-Mar-08	12,532.60
24-Mar-08	12,548.64
20-Mar-08	12,361.32
19-Mar-08	12,099.66
18-Mar-08	12,392.66
17-Mar-08	11,972.25
14-Mar-08	11,951.09
13-Mar-08	12,145.74
12-Mar-08	12,110.24
11-Mar-08	12,156.81
10-Mar-08	11,740.15
7-Mar-08	11,893.69
6-Mar-08	12,040.39
5-Mar-08	12,254.99
4-Mar-08	12,213.80
3-Mar-08	12,258.90

Sketch your scatter plot here…

(1) Do you see any patterns in the above data?

(2) Find the linear regression equation and r-value…

_____

(3) Type the equation in Y_1, or use the easy way that I showed you before and graph the model over the data.

(4) Do you think this model will be accurate for Interpolation? Extrapolation? Explain.

(5) Based on your experience of the stock market, do you think a non-linear model would work better?

Associated End of Booklet exercises are 22 – 25

Topical Summary of Linear Functions

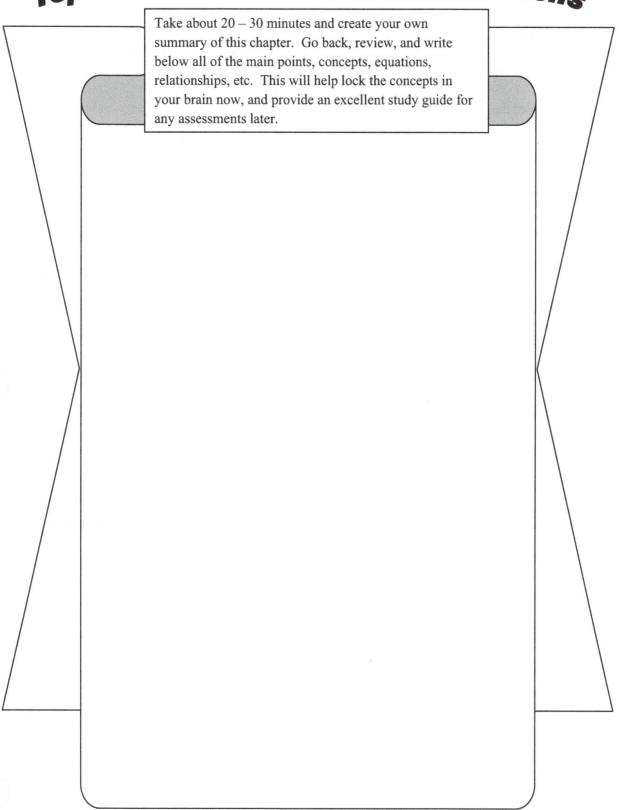

Take about 20 – 30 minutes and create your own summary of this chapter. Go back, review, and write below all of the main points, concepts, equations, relationships, etc. This will help lock the concepts in your brain now, and provide an excellent study guide for any assessments later.

End of Booklet Exercises

Übung macht den Meister! ((Translation "Practice makes perfect!"))

1) Take out a sheet of notebook paper, draw the x-y axes on it so it fills the entire paper, and label quadrants 1 through 4 in their appropriate places. Then sign the name of the guy down the hall and tack it up to your RA's door.

2) For each of the following points, write either Q1, Q2, Q3, Q4, x-axis, y-axis, or origin, depending on where the point lies.

(a) $(3, 6)$ (b) $(-3, 6)$ (c) $(3, -6)$ (d) $(-3, -6)$

(e) $(3, 0)$ (f) $(0, 3)$ (g) $(-3, 0)$ (h) $(0, -3)$

(i) $(0, 0)$ (j) $(.001, .001)$ (k) (π, e)

3) Choose a value for x and/or y to make the following statements true. If the statement cannot be made true, state so.

(a) $(5, y)$ with the point in Q1   (b) $(5, y)$ with the point in Q3

(c) $(x, 5)$ with the point in Q2   (d) $(x, 5)$ with the point in Q3

(e) $(5, y)$ with the point on the x-axis   (f) $(5, y)$ with the point on the y-axis

(g) $(5, y)$ with the point at the origin   (h) (x, y) with the point at the origin

4) Find the slope of the line that passes through each pair of points below.

     (a) $(1, 2)$ & $(3, 6)$ (b) $(-5, 10)$ & $(4, -8)$

     (c) $(0, 0)$ & $(2, 9)$ (d) $(2.5, 4)$ & $(-12, 4)$

(e) (3, -1) & (3, 13)

(f) (.25, 4.67) & (-3.12, 9.06)

(g) (-117, 38) & (56, -92)

(h) (1/2, 2/3) & (4/7, -9/5)

(i) (5, y) & (7, -3)

(j) (π, 1) & (0, π)

(k) (a, b) & (c, d)

(l) ($, @) & (*, ^)

5) Use the Distance formula to find the distance between the points in questions 4 (a) – (f).

(a)

(b)

(c)

(d)

(e)

(f)

6) Estimate the x and y intercepts of the following graphs. Write each intercept in (x, y) form.

(a)

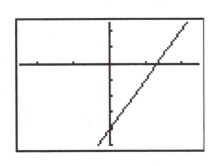

x-intercept _____

y-intercept _____

(b)

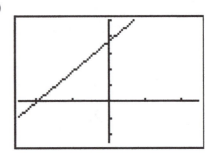

x-intercept _____

y-intercept _____

(c)

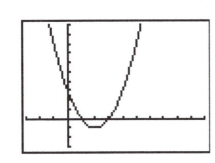

x-intercept _____

y-intercept _____

(d)

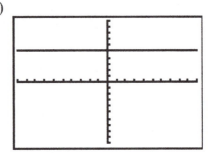

x-intercept _____

y-intercept _____

7) Identify the slope and y-intercept (that's, *m* and *b*) for each of the linear equations below. You may need to change them into $y = mx + b$ form first.

(a) $3y = 6x - 7$

(b) $4f(x) = 7 - 2x$

(c) $Z(p) = -2p + 1$

(d) $4y + 7x - 10 = 0$

(e) $2 - x = -y$

(f) $y = \dfrac{x - 2}{3}$

8) Your friend says that her test scores have steadily increased by 15 points per exam, starting at a low of 58. This relationship is linear. Identify the slope and y-intercept of the relationship.

9) Your grandmother started her own business recently. She is downloading songs into people's ipods™ for them. She charges $8 as a set-up fee and $0.05 per song downloaded. Write a function that relates her income (I) as a function of the number of songs downloaded (s).

10) In the *Booklet on Real Numbers*, we introduced the Western Chromatic Scale. Starting with any natural note on a piano, to get to that same note one octave higher, you would move7 white keys to the right. To get to that same note one octave lower, you would move7 white keys to the left. Write a function that has the number of octaves changed as the input (N) and the amount of keys moved as the output (k). What is the slope and y-intercept of this function?

11) Given the function $f(x) = -7x - 5$. Answer the following questions…

(a) Write two functions that are parallel to the given function.

   $g(x) =$ _____ $h(x) =$ _____

(b) Write two functions that are perpendicular to the given function.

   $g(x) =$ _____ $h(x) =$ _____

12) Write one equation of a line that is ∥ to $6y - 4 = 9x$

13) Write one equation of a line that is⊥ to $x - y = 7$

14) In question 4) you found the slope of the line that passes through each pair of points below. Use your answers to question 4) and find the equation of each line.
    (a) (1, 2) & (3, 6)   (b) (-5, 10) & (4, -8)

(c) (0, 0) & (2, 9)   (d) (2.5, 4) & (-12, 4)

(e) (3, -1) & (3, 13)   (f) (.25, 4.67) & (-3.12, 9.06)

(g) (-117, 38) & (56, -92)   (h) (1/2, 2/3) & (4/7, -9/5)

15) Find the equation of the line that passes through the point (0, 4) and is parallel to the line $y = -9x - 15$.

16) Find the equation of the line that passes through the point (0, -1) and is perpendicular to the line $y = -9x - 15$.

17) Find the equation of the line that passes through the point (2, 5) and is perpendicular to the line $y = x + 8$.

18) Find the equation of the line that passes through the point (-3, 11) and is parallel to the line $y = 5x$.

19) Find the equation of the line that passes through the point (1.25, -3.75) and is perpendicular to the y-axis.

0) The amount of miles a family on vacation can travel in their vehicle in one day is a function of the number of hours on the road. What is a practical domain and range for this situation?

21) An object is thrown from the Empire State building. The distance the object is from the ground is a function of time. Given that the building is about 1250 feet, what might be a practical domain and range for this situation? (Note: Do not attempt this on your own…make sure an adult is with you).

22) The National Association of Realtors puts out reams of data regarding home sales. The first table below is the average number of homes sold in the United States for a given year, and the table below that is the average sales price for U.S. homes.

Year	Homes sold in the US
2004	7,100,000
2005	7,076,000
2006	6,478,000
2007	5,652,000

TABLE 1

Year	Average sales price ($)
2004	185,200
2005	219,600
2006	221,900
2007	219,000

TABLE 2

(a) Using the years 2005 and 2006, find the equation of the line that represents the number of homes sold as a function of year.

(b) Use your TI-83 to calculate the line of best fit for the data in table 1.

(c) How close is your equation from part (a) to the LOBF in part (b)? How can you account for the differences?

(d) Using the years 2005 and 2006, find the equation of the line that represents the average sales price as a function of year.

(e) Use your TI-83 to calculate the line of best fit for the data in table 2.

(f) How close is your equation from part (d) to the LOBF in part (e)? How can you account for the differences?

23) The US Bureau of the Census provides the following average sales prices for new construction homes from 1970 to 2005

Year	Average sales price ($)
1970	26,600
1975	42,600
1980	76,400
1985	100,800
1990	149,800
1995	158,700
2000	207,000
2005	297,000

(a) Create a scatter plot of this data with your TI-83 and sketch it above.

(b) Have your calculator find the LOBF for this data. Write the equation and the r-value. What does the r-value tell you about the match between the equation and the data?

(c) Use the regression equation (LOBF) to estimate the cost of a new construction home in 2008.

(d) Use the internet to research the actual cost of a new construction home in the US for 2008 and discuss the difference, if any.

24) The Department of Labor lists the following Federal Minimum Wages in the US for certain years. Note that the wage for 2009 is anticipated at the writing of this text.

Year	Minimum Wage ($)
1938	.25
1939	.30
1945	.40
1950	.75
1956	1
1961	1.15
1963	1.25
1990	3.8
1991	4.25
1996	4.75
1997	5.15
2007	5.85
2008	6.55
2009	7.25

(a) Use your TI-83 to find the LOBF from 1938 to 1963, and the LOBF from 1990 to 2009. Remember to make both 1938 and 1990 year 0. Write them below.

(b) Compare the slopes of the two lines. Based on the slopes, what statement can you make about changes in minimum wages earlier in the 20th Century compared to later?

25) *Paying for College*

The following graphic contains the average tuition and required fees for full-time students at private, 4-year colleges over the last two decades.

YEAR	COST ($)
1996	12,990
1994	11,480
1992	10,290
1990	11,380
1989	10,350
1979	3,810

(*i*) Let t represent the number of years since 1979. Create a scatter plot of your data and sketch it below.

(*ii*) Does there appear to be a linear relationship between the year and the average cost of tuition? Explain

(*iii*) Estimate the line of best fit. Go back to your sketch and draw the line that will have as many data points as close to the line as possible.

(*iv*) Estimate the slope of your line. What is the practical meaning of the slope in this situation?

(*v*) What is the vertical intercept of your line of best fit? Does this number have any practical meaning in the situation?

(*vi*) Using your answers to parts (*iv*) and (*v*) find the equation of your line of best fit. Use the form $y = f(x) = mx + b$.

(*vii*) To measure the "goodness of fit" of your line, complete the following table.

Input (t)	Actual Output	Your model's output	|col. 2 – col. 3|
0	3810		
10	10,350		
11	11,380		
13	10,290		
15	11,480		
17	12,990		

(*viii*) Determine the sum of the values in the last column. This is the *error* of your line. The smaller the error; the better the fit.

(*ix*) Use your TI-83 to calculate the regression equation

(*x*) Now use the regression equation to model the data. Find the error below

Input (t)	Actual Output	Your model's output	|col. 2 – col. 3|
0	3810		
10	10,350		
11	11,380		
13	10,290		
15	11,480		
17	12,990		

(*xi*) Calculate the sum and compare the 2 errors. Which is better?

(*xii*) Use the regression equation to predict the tuition in 2005? Is this a good estimate? Why or why not?

Booklet on Several Linear Families

Also called "Systems" of Linear Equations

One line is boring! Two lines are also boring except that we notice one very interesting point. Can anyone guess what that one point is?

If not, try this. On the space below, draw <u>any</u> two lines with different slopes (i.e., not parallel). Do you notice that whatever two lines you draw, they always do something?

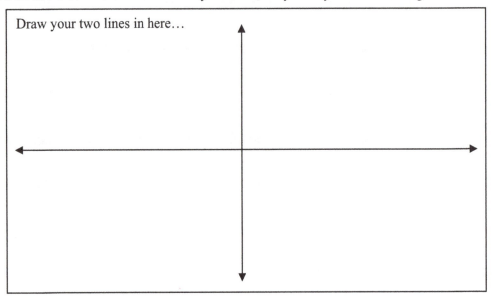

Draw your two lines in here…

Intersect! The interesting point is where these lines meet (cross, intersect, etc.). We call this point the <u>intersection point</u>, or more frequently, the <u>solution</u> of the system.   Since both lines meet at that one point (and one point only), that specific point *satisfies* both equations. In other words, the point (or solution) makes both equations true! You might say 'that's common sense!'…well, maybe… but because it is true it allows us to do some good things algebraically.

Intersection point

This point satisfies both linear functions. That is, it makes both equations *true* when x and y are plugged back in.

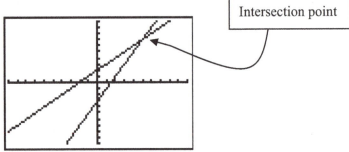

Many, many textbooks teach several different algebraic methods to solve systems, but I'm only going to show you one way (algebraically) because it always works! Just like $y = mx + b$ is the only form of a line that you'll ever need, so the **substitution method** is the only way to go when solving systems. And, because you have had a lot of practice putting linear equations in $y = mx + b$ form, this method will come easy to you, I promise!

Substitution always works!

The Substitution Method

Substitution relies on the following principle (called Transitivity):

"If a = b, and b = c, then a = c"

An easy example of this is money. Since four quarters equals a dollar, and ten dimes also equals a dollar, then four quarters must also equal ten dimes.

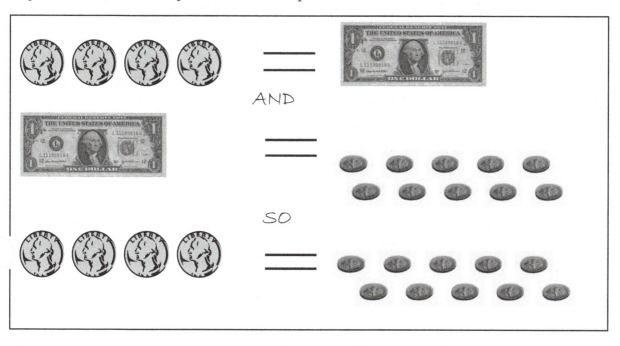

Let's apply this to lines. How does this apply to lines?...If y equals one mx + b and y equals another mx + b, then the mx + b parts must also be equal.

$$y = m_1x + b_1 \quad \text{and} \quad y = m_2x + b_2$$

a = b and a = c

Then

$$m_1x + b_1 = m_2x + b_2$$

b = c

We eliminate the y's and set the "mx + b" parts equal to each other. This allows us to solve for x because it is the only variable. Then we can substitute back into either equation to find y.

Example:

We want to find the solution to the following system of linear equations:
$$y = -x + 2 \quad \& \quad y = -2x + 4$$

Let's apply transitivity. Since y (*one dollar*) equals "-x + 2" (*four quarters*) and y (*one dollar*) **also** equals "-2x + 4" (*ten dimes*) then –x + 2 (*four quarters*) must equal –2x + 4 (*ten dimes*). In other words, we are "substituting" for y (and thereby eliminating it).

$$
\begin{array}{ll}
-x + 2 = -2x + 4 & \text{(now we can solve for x)} \\
\underline{+2x \qquad +2x} & \text{(add 2x to both sides)} \\
x + 2 = 4 & \text{(subtract 2 from both sides)} \\
\underline{-2 \quad -2} & \\
x = 2 & \text{(the x-coordinate of the solution)}
\end{array}
$$

Once we have the x-coordinate of the solution, we can find y by substituting x in to **either** of the two original equations. (Do you remember why either one works??)

$$y = -(2) + 2$$
$$y = 0$$

Our solution to the system is the point (2, 0). This is where the two lines intersect.

There is one condition to using substitution. Both equations **must** be in $y = mx + b$ form. The reason is simple: -3y does not equal y, only y equals y. So, one side of both equations must simply have y all by itself. We guarantee this by putting both equations in $mx + b$ form.

Example:

We want to find the solution to the following system of linear equations:

$$2x + 5y = 1 \quad \& \quad 10 - 4x = 3y$$

Let's put both into $y = mx + b$ form first. Follow these to see if you understand

$$
\begin{array}{ll}
\text{(subtract 2x from both sides)} \quad 2x + 5y = 1 & \qquad \dfrac{10 - 4x}{3} = \dfrac{3y}{3} \quad \dots\text{divide both sides by 3} \\
\qquad\qquad \underline{-2x \qquad\quad -2x} & \\
\qquad\qquad \dfrac{5y}{5} = \dfrac{-2x}{5} + \dfrac{1}{5} & \qquad \dfrac{10 - 4x}{3} = y \quad \dots\text{separate the fraction} \\
\text{(divide both sides by 5)} &
\end{array}
$$

$$y = -2/5 \, x + 1/5 \quad \text{and} \quad y = -4/3 \, x + 10/3 \quad \dots\text{done}$$

Let's change both to decimal form:

$$y = -0.4x + 0.2 \quad \text{and} \quad y = -1.33x + 3.33$$

Now we can apply substitution:

one y   =   the other y

-0.4x + 0.2 = -1.33x + 3.33   (get the x's one side and the numbers on the other)

$$\begin{array}{ccccc} -0.4x & + & 0.2 & = & -1.33x & + & 3.33 \\ +1.33x & & -0.2 & & +1.33x & & -0.2 \\ \hline 0.93x & & & = & & & 3.13 \\ 0.93 & & & & & & 0.93 \end{array}$$

x ~ 3.37   (The tilde "~" means approximately)

Substitute back to find y:

y = -0.4(3.37) + 0.2   so y ~ -1.15

Our solution is (approximately) the point (3.37, -1.15)

If we would have stayed with fractions we could have arrived at an exact answer. In either case the answer is much better than anything we could have done by just looking at a graph and guessing.

Did you remember to Check??

Since we are saying that the above point satisfies **both** equations, we should check **both** equations. They both must be satisfied in order for the solution to be correct.

y = -0.4x + 0.2   and   y = -1.33x + 3.33

-1.15 = -0.4(3.37) + 0.2   and   -1.15 = -1.33(3.37) + 3.33

-1.15 = -1.15 √   and   -1.15 = -1.15   √

Professor's Practice Pause

(1) Circle the point of intersection on the graph below and write an estimate of the solution.

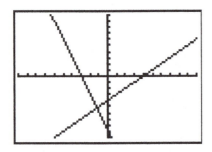

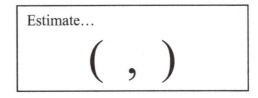

Estimate…

(,)

(2) Now let's find the solution exactly using the substitution method. The equations graphed above are $f(x) = -3x - 9$ and $f(x) = x - 4$.   {Remember y is the same thing as f(x)} Be sure to check your solution back in both lines!

(3) Find the solution to the following system of linear equations. Put both equations in y = mx + b form and use substitution.

$$-3y = 5x - 7$$
$$2y + 6x = 10$$

Associated End of Booklet exercises are 1 – 9

Ye Olde Lemonade Stand...

Let's say we can determine linear profit functions for two summer business ventures that you are choosing between (lemonade stands). In each equation, the total *daily* profit (P) is a function of the cost per lemonade (c) and your start-up costs (the y-intercept value).

For Lemonade stand 1, you charge $0.95 a lemonade, but you have to buy lemons, cups, sugar, etc. at a total cost of $15 a day.

For Lemonade stand 2, you charge only $0.75 a lemonade, but you "cheat" by buying pre-packaged mix instead of squeezing the lemons. The mix costs $11.50

Let's find the following things:
a) The daily profit equation for each stand.
b) The number of lemonades it takes to reach our break-even point for each stand.
c) The number of lemonades sold so that the daily profits for each stand are the same.

a) Lemonade Stand 1:
Daily profit = "price per lemonade" times "number of lemonades old" - "start-up costs"
$$P_1(x) = 0.95x - 15.$$

Lemonade Stand 2:
Daily profit = "price per lemonade" times "number of lemonades old" - "start-up costs"
$$P_2(x) = 0.75x - 11.50$$

b) The break-even point for each day is when Profit is zero $(P(x) = 0)$
#1 $\quad 0 = 0.95x - 15$
$\qquad 15 = 0.95x$
$\qquad x = 15 / 0.95 = 15.8$ or $\underline{16\ lemonades}$ to break even

#2 $\quad 0 = 0.75x - 11.50$
$\qquad 11.50 = 0.75x$
$\qquad x = 11.5 / 0.75 = 15.3$ or again about $\underline{16\ lemonades}$.

Looks like both businesses will make about the same money each day. Or will they??

c) Solve the system. In other words, when are they equal?
$$P_2(x) \ = \ P_1(x)$$
When does $\quad 0.75x - 11.5 = 0.95x - 15$? .....subtract 0.75x from both sides
$\qquad\qquad\qquad -11.5 = 0.2x - 15 \quad$ ......add 15 to both sides
$\qquad\qquad\qquad\quad 3.5 = 0.2x \qquad\qquad$ ......divide both sides by 0.2
$\qquad\qquad\qquad 17.5 = x$

So the lemonade stands receive equal daily profits after about $\underline{18\ lemonades}$. But what happens after 18 lemonades? Which one makes more money?

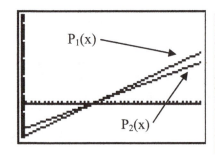

After 18 lemonades, $P_1(x)$ always dominates $P_2(x)$, because the slope of $P_1(x)$ is greater than the slope of $P_2(x)$.

If you want to make money, choose the first lemonade stand (and sell more than 18 lemonades a day!)

Different types of systems

If you've studied this topic before, you know that most textbooks name these different types of systems as *Consistent, Inconsistent,* and *Dependent*. While those are apt names, more revealing titles might be **Intersecting, Parallel**, and *Algebraically Disguised Imposter*…ok, maybe that's over the top…how about **Equal**. An intersecting system has two lines that intersect, giving 1 solution. A parallel system has two parallel lines, which do not intersect, giving no solutions. An equal system has two lines that are the same, which intersect in all points, giving an infinite number of solutions.

System	Types of lines	Solutions	Why?
Intersecting	Two lines with different slopes	1	The lines only meet at one point
Parallel	Two lines with the same slope	0	The lines do not meet at all…ever!
Equal	Two lines with the same equation	∞	The lines touch at every point.

TI-83 examples of these three systems are shown below…

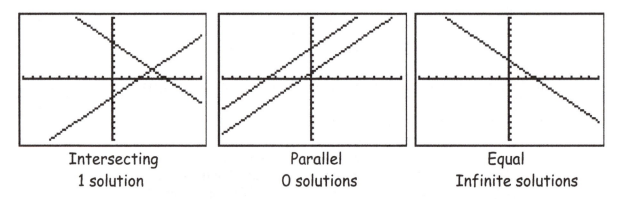

Intersecting	Parallel	Equal
1 solution	0 solutions	Infinite solutions

How do we identify different systems? That's easy…**y = mx + b**. By the way, did I tell you that y = mx + b does it all?? If you put both equations in the system in y = mx + b form, you can immediately see if they have different slopes, the same slope, or the same equation.

Example: Identify the type of system given.

(a) $3y = x - 5$
     $y + 10 = 7x$

(b) $2x - y = 4$
     $4x + 2y = -8$

(c) $y = 4x - 1$
     $3y - 6 = 12x$

Solution: Put all equations in $y = mx + b$ form, if they aren't already.

(a) $\underline{3y} = \underline{x - 5}$
    $\ \ 3 \ \ \ \ \ 3$

$\underline{y = (1/3)x - 5/3}$

$y + 10 = 7x$
$\underline{-10 \ \ \ \ -10}$

$\underline{y = 7x - 10}$

(b) $-2x - y = 4$
    $\underline{+2x \ \ \ \ \ \ \ \ +2x}$
    $\underline{-y} = \underline{2x + 4}$
    $-1 \ \ \ \ \ \ -1$
    $\underline{y = -2x - 4}$

$4x + 2y = -8$
$\underline{-4x \ \ \ \ \ \ \ \ -4x}$
$\underline{2y} = \underline{-4x - 8}$
$\ 2 \ \ \ \ \ \ \ \ \ 2$
$\underline{y = -2x - 4}$

(c) $\underline{y = 4x - 1}$ (done)

$3y - 6 = 12x$
$\underline{+6 \ \ \ \ \ +6}$
$3y = 12x + 6$

$\underline{3y} = \underline{12x + 6}$
$\ 3 \ \ \ \ \ \ \ 3$

$\underline{y = 4x + 2}$

(a) This system is Intersecting, because the lines have different slopes.

(b) This system is Equal because the equations are the same.

(c) This system is Parallel because the lines have the same slope (but different intercepts).

Professor's Practice Pause

For questions (1) and (2), write the two equations for each system by inspection. That is, just using the information given in the scenario. Do not solve the systems.

(1) *Music Matez* offers guitar lessons for $25 per hour with an initial sign-up fee of $50. *Larry's Lessons* waves the initial fee, but charges $32.50 per hour for each lesson. Write equations for the total cost of lessons in both places as a function of number of hours.

(2) *Verderber's Virtuosos*, a private, music school in New England, charges $2200 per credit hour and about $10,500 for room, board, texts, and fees. The *Snare Symposium*, a not-so-well-known music school in downtown Schenectady charges $1025 per credit hour and only $800 for all texts and fees (since it is a commuter college). Write equations for the total cost of attending these fine, disreputable institutions as a function of credit hours taken.

(3) Two linear functions form a system. The first line is $f(x) = -3x + 5$. Write a possible function for the second line if the system is:

(a) Intersecting, $g(x) =$ _____

(b) Parallel, $g(x) =$ _____

(c) Equal, $g(x) =$ _____

(d) $\perp$ to f(x), $g(x) =$ _____

(4) The following table shows the Manufacturer's Suggested Retail Price (MSRP) and the annual depreciation value of two new cars; the Volkswagen Jetta and the Kia Spectra.

(a) Write linear functions for the value of each car, V, as a function of number of years, x.
(b) After how many years will the cars have the same value?

Model	MSRP ($)	Depreciation per year ($)
VW Jetta	18,750	1125
Kia Spectra	14,680	1490

Associated End of Booklet exercises are 10 – 19

The 6 magic buttons (revisited)....The Intersect Method

In the *Booklet on Equations and Expressions*, we looked at a TI technique called the "6 magic buttons." If you need to, pull out that Booklet and take a quick look to refresh your memory.

We can use this technique to solve systems of linear equations graphically. In our substituted equation (after eliminating the y's) we have $mx + b = mx + b$. We put the left hand side in Y_1, the right hand side in Y_2 and press **2nd, TRACE, 5, ENTER, ENTER, ENTER**.

Let's go back to the first example... $-x + 2 = -2x + 4$

  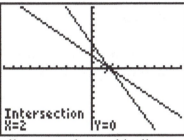

It's always a good thing when you get the same answer algebraically as you do graphically, right?

And another...the second example... $y = (-2/5)x + 1/5$ & $y = (-4/3)x + 10/3$

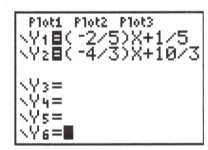  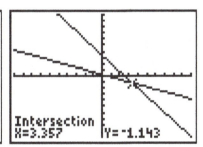

Our answer was (3.37, -1.15), which is close considering we used rounded fractions.

Use the intersect method to check the answer you got in the Jetta vs. Spectra comparison in the last Professor's Practice Pause. Sketch the two lines on the TI plot below and label your axes with reasonable values (based on your window).

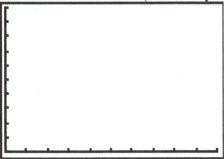

Group Pause

In groups of two or three, solve this system algebraically using the method of substitution, and then check your answers using the Intersect Method on the TI.

You are buying vanilla beans for your new business, "The Linear Lunchbox." You discover that low grade beans sell for $0.80 a pound, and high grade beans sell for $1.45 a pound. To meet the demands of the upcoming month, you need to buy a total of 5000 pounds of beans. You must spend the remainder of your bean money... $6, 500.

Let "x" be the number of pounds of low grade beans and let "y" be the number of pounds of high grade beans that you buy.

Write a system of linear equations (two) that describe the situation above and solve for the weight of each grade bean that you need to buy.

Another Group Pause

What does this mean?

(4) *Straight-line (linear) depreciation*

Straight-line (or linear) depreciation is the most common method used to *depreciate* assets for tax purposes. These assets can be houses, boats, machinery, and cars. The formula for linear depreciation looks like this:

$$\text{Annual depreciation expense} = \frac{\text{Cost of asset} - \text{Scrap value}}{\text{life span (in years)}}$$

For a new car, the cost is your purchase price, the scrap value can be considered your approximated trade-in value, and the lifespan is the number of years you own before you trade it in. The table below compares these aspects for two new vehicles – Honda Accord and Lexus RX350.

Model	Purchase Price ($)	Number of years of use	Scrap Value ($)
Honda Accord	26,000	10	2350
Lexus RX 350	37,500	10	4825

(a) Calculate the Annual Depreciation expense for both vehicles.

(b) Write linear functions for the value of each car, V, as a function of number of years, t. Assume the depreciation remains constant.

(c) Use the Intersect method to find how many years of use are required for the values to become equal? Write your answer below and explain the *practical* significance of your answer.

Associated End of Booklet exercise is 20

Solving 3 equations with 3 unknowns

Very Important Point #8 (VIP 8)…

"In order to solve for *n* different variables, you need *n* different, independent equations that contain those variables."

Think about what we've been doing so far…solving for x and y. That's 2 unknowns. To do that, we had to use 2 equations, both containing x and y. If we wanted to solve for 3 variables, say x, y, and z, we would need three, different (*independent*) equations with those 3 variables. Independent equations are actually different equations…they aren't the same equation disguised by algebra.

Consider these two equations…

$$y = 3x - 8 \ \& \ 6x - 2y = 16$$

To the ordinary observer, they look quite different. However, the one on the right is an algebraically disguised imposter! We shall prove it by putting this equation in $y = mx + b$ form, perhaps one of the greatest things ever invented (or discovered?)

RIGHT BRAIN / LEFT BRAIN

Was math invented or discovered? This is a question that has intrigued both student and professor alike for many centuries. Some say that math was discovered. In other words, the principles were always there, it just took someone to find them. Others would say that math was invented; that is, people created mathematical relationships that were never there before from some already known ideas. Still others would argue that it's a little bit of both.

What's your opinion? We'd like to know…

Back to our algebraically disguised imposter… 6x – 2y = 16

$$\underline{\quad -6x \qquad \quad -6x \quad} \ \text{…subtract 6x from both sides}$$
$$-2y = -6x + 16$$

$$\frac{-2y}{-2} = \frac{-6x + 16}{-2} \quad \text{….divide both sides by -2}$$

$$y = -6x/-2 + 16/-2 = \mathbf{3x - 8}$$

They are exactly the same equations, just in different forms. We call these *dependent* (or Equal) equations or a dependent (or Equal) system. That's why in order to solve systems for *n* variables, we need n independent equations. If 2 equations were equal they could only count as 1 unique equation, not 2, and so we could not solve our system.

Let's look at an equation in three variables…for example: $3x - 4y + 7z = 1$

We can't put this equation into $y = mx + b$ form since there is also a z term. How do we even know that the equation is linear? Well, If we wanted to we could put it into $y = mx + nz + b$ form, which suggests that the equation is linear in 3-dimensions. But that doesn't help us use the method of substitution to solve these 3 x 3 systems (read *three by three systems*). In other words there are 3 equations and 3 unknowns.

Here's our strategy…we will use substitution again to turn the 3 x 3 into a 2 x 2 by eliminating one variable. Once we're there we know how to proceed (that's what we've been doing for the past 10 pages!!) It looks like this…

---

3 x 3 $\rightarrow$ 2 x 2 $\rightarrow$ 1 x 1 $\rightarrow$ solve for variable. Then we substitute back into previous equations to find the other two variables. Finally, we check our answers in all 3 original equations.

Example: Solve the following 3 x 3 system of linear equations

(a) $x + 2y - 3z = 5$
(b) $-x + y + 2z = 0$
(c) $2x - y + z = -1$

Important note: In $y = mx + nz + b$ form, these equations are…
(a) $y = -(1/2)x + (3/2) z + 5/2$
(b) $y = x - 2z$
(c) $y = 2x + z + 1$
No equation is a constant multiple of another equation. They are all independent, which means there is a solution (x, y, z)

There are many, many, many ways to solve these types of systems. We will just pick one way and stick with it!

Let's solve equation (b) for y and then substitute that into equations (a) and (c), thereby eliminating the y variable.

$y = x - 2z$ (we did this already in the Important note above)

So equation (a) becomes
$x + 2(x - 2z) - 3z = 5$
$x + 2x - 4z - 3z = 5$
$3x - 7z = 5$….we now have 1 equation with x and z.

Equation (c) becomes
$2x - (x - 2z) + z = -1$
$2x - x + 2z + z = -1$
$x + 3z = -1$….we now have a second equation with x and z

We have just went from a 3 x 3 to a 2 x 2

We can now use either substitution or the Intersect method to solve our 2 x 2. Let's continue with an algebraic solution by changing both into "x = mz + b" form and using substitution.

(a) $3x - 7z = 5 \rightarrow 3x = 7z + 5 \rightarrow \underline{x = (7/3)z + 5/3}$

(c) $x + 3z = -1 \rightarrow \underline{x = -3z - 1}$

Set the two equations equal to each other and solve...

$$(7/3)z + 5/3 = -3z - 1 \dots\dots\text{This is our 1 x 1}$$
$$7z + 5 = -9z - 3 \quad \dots\dots\text{multiply both sides by 3 to clear fractions}$$
$$\underline{+9z \qquad +9z} \qquad \dots\dots\text{Add 9z to both sides}$$
$$16z + 5 = -3$$
$$\underline{\quad -5 \quad -5} \qquad \dots\dots\text{subtract 5 from both sides}$$
$$\frac{16z}{16} = \frac{-8}{16} \qquad \dots\dots\text{divide both sides by 16}$$

$$\underline{z = -1/2}$$

Now we choose any equation to back-substitute and find another variable

Choose $x = -3z - 1$. Since $z = -1/2$, we have $x = -3(-1/2) - 1 = 3/2 - 1 = \frac{1}{2}$. $\underline{x = 1/2}$

And once more.....

In the original equation (b) we know that $y = x - 2z$. $y = 1/2 - 2(-1/2) = 1/2 + 1 = 3/2$. $\underline{y = 3/2}$

Our solution is the point $(x, y, z) = \underline{(1/2, 3/2, -1/2)}\dots\text{or}\dots(0.5, 1.5, -0.5)$

Now to check in all 3 original equations

(a) $x + 2y - 3z = 5$ $1/2 + 2(3/2) - 3(-1/2) = 5$ $\frac{1}{2} + 6/2 + 3/2 = 5$ $5 = 5\dots\checkmark$

(b) $-x + y + 2z = 0$ $-1/2 + 3/2 + 2(-1/2) = 0$ $1 - 1 = 0$ $0 = 0\dots\checkmark$

(c) $2x - y + z = -1$ $2(1/2) - 3/2 + (-1/2) = -1$ $1 - 3/2 - \frac{1}{2} = -1$ $-1 = -1\dots\checkmark$

Afterthoughts...

(1) These problems are long! Take your time, work neatly, label each equation, and write out the step that you're in, as I did above using the **Comic Sans MS** font.

(2) There are many ways to solve 3 x 3's. I chose to substitute, and hence eliminate, y. I could have chosen to eliminate x or z…it doesn't matter. Choose a path and stick with it.

(3) Always check your answers! Your 3-D point (x, y, z) must work in all 3 original equations. If not, it's not a solution and it means you need to go back and look at your work.

(4) Your professor will probably end up showing you the Addition method as well. That's ok, either way works. My goal is to present one algebraic method (substitution) that can always be used for linear functions and systems of linear functions. I believe this is less confusing *por vous*!

Application

Your friend is buying candy for his sweetie for Valentine's Day. He decides to buy chocolate-covered almonds, peanut butter cups, and toffee crunch. The chocolate-covered almonds cost $4.50 per pound, the peanut butter cups are $1.50 per pound (he got those at Walmart), and the toffee crunch costs $3.75 per pound. Your friend decides three things:

1. He is going to spend $35.
2. He is going to get his sweetie 12 pounds of candy.
3. He is going to buy 3 more pounds of peanut butter cups than chocolate almonds.

If **a**, **p**, and **t** represent the weights of almonds, peanut butter cups, and toffee (respectively), *write* <u>three</u> equations based on your friend's three decisions above and solve for the weights.

First we need 3 equations. Each of the three stipulations above will provide one equation.

(1) $4.5a + 1.5p + 3.75t = 35$

> The money equation. Notice that all the terms of the equation have units of $. Price per pound times number of pounds always give price….($/lbs)*lbs = $

(2) $a + p + t = 12$

> The pounds equation. Notice that all the terms of the equation have units of pounds.

(3) $p = a + 3$

> Equation 3 is also a pounds equation. Remember that the English phrase "three more *than* x" means x + 3

Changing the 3 x 3 into a 2 x 2…

Let's use equation (3) to substitute (eliminate) the variable p in both equations (1) and (2). Equation (3) is the least complex one – typically those are the ones used to substitute.

(1) $4.5a + 1.5\mathbf{(a + 3)} + 3.75t = 35$

    $4.5a + 1.5 + 4.5 + 3.75t = 35$

      $6a + 3.75t + 4.5 = 35$

             $-4.5 \quad -4.5$ ……..subtract 4.5 from each side

        $\underline{6a + 3.75t = 30.5}$

(2) $a + p + t = 12$

    $a + \mathbf{(a + 3)} + t = 12$

      $2a + t + 3 = 12$

      $\underline{2a + t = 9}$ ……subtracted 3 from both sides

Now we solve our 2 x 2

To shake things up a bit, let's use the Intersect method on the TI-83. In order to do that, we have to change our variables to x and y (not a and t). So let **x be t**, and **y be a**…

(1) $6y + 3.75x = 30.5$ &   (2) $2y + x = 9$

    $6y = 30.5 - 3.75x$ $2y = 9 - x$

    $y = (30.5 - 3.75x)/6$ $y = (9 - x)/2$

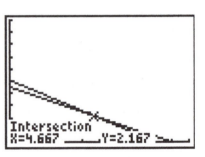

This tells me that x (really t) = **4.667 lbs**, and that y (really a) = **2.167 lbs**.

Now we can find p by going back to equation (3) $p = a + 3 = 2.167 + 3 = \mathbf{5.167\ lbs}$.

Your friend is buying 4.667 lbs of toffee, 2.167 lbs of almonds, and 5.167 lbs of peanut butter crunch. He's also buying stock in the ADA (American Dental Association)!!

Checks.....

(1) $4.5a + 1.5p + 3.75t = 35$

    $4.5(2.167) + 1.5(5.167) + 3.75(4.667) = 35 \;\rightarrow\; 35 = 35$….√

(2) $a + p + t = 12$

      $2.167 + 5.167 + 4.667 = 12.. \rightarrow\; 12 = 12$…√

(3) We know equation 3 works since we used it to find p….√

Professor's Practice Pause

Given the following 3 x 3 system of linear equations…

$$3\,x + 2\,y + z = 1$$
$$x + y - 2\,z = -4$$
$$2\,x - 3\,y + 3\,z = 1$$

(a) Place all three equations in $y = mx + nz + b$ form to see if the system is *independent*.

(b) If the system is independent, find the solution using substitution, or substitution combined with the Intersect method.

Associated End of Booklet exercises are 21 and 22

Topical Summary of Linear Systems

Take about 20 – 30 minutes and create your own summary of this chapter. Go back, review, and write below all of the main points, concepts, equations, relationships, etc. This will help lock the concepts in your brain now, and provide an excellent study guide for any assessments later.

A featured guest in mathematical history…

Hypatia of Alexandria

I present you with a synopsis of the life of Hypatia.

Hypatia (pronounced Hip' a tee' ya) was a Greek, female mathematician around 300 BC. If any of you know about Greek culture, the female of the species is not regarded as highly as the male - more so back then. To make matters worse, she was smarter (mathematically) than just about everyone on the block (including the guys). This fact really got their knickers in a twist. So much so, that they despised her enough to seek her demise. Unfortunately for them, Hypatia's father was the governor at the time.

Well, when daddy died, the good ol' boys done did 'er in!! The mere fact that they executed her is bad, but it is the method wherein the depravity of the human mind sheds its opaqueness.

While she was walking the main thoroughfare, they grabbed hold of her, threw her on a chariot and brought her to the temple. On the altar, they stripped her of her clothes, and scraped the flesh from her bones with sharp oyster shells. Yikes!!

Much of mathematics has come to us via the blood of its founders. Many people throughout the history of mathematics have faced persecution, ridicule, and death to bring us what we now know about it.

While it is sometimes preferable to stand on the shoulders of giants and gaze upward into the heavens, it is far better to construct the foundation of the mathematical tower from the ground up, brick by brick.

Hypatia takes a bow to all of you mathematical masons.

Google "Hypatia of Alexandria" for some more insight into her work and life.

End of Booklet Exercises (If it's not getting better, spray Windex on it!)

1) What point is the *solution* to a system of linear equations?

2) If A = B and B = C, then…?

3) If y = 2x and y = -x, then 2x = _____ ?

4) You buy $100 worth of triple cheeseburgers at $2.95 each. *Estimate* the number of cheeseburgers you can purchase. Calculate the actual number of cheeseburgers you can buy. Is your estimate close?

5) Estimate the solution for the following graphs of systems of linear equations.

(a)
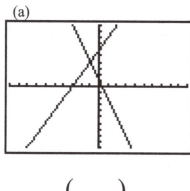

(b)
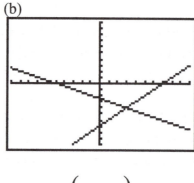

(c)
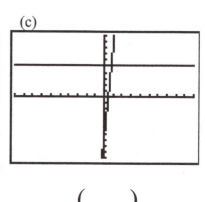

(,)　　　　　　(,)　　　　　　(,)

6) The systems below correspond to the graphs above. Use the method of substitution to solve for the solution exactly.

(a) y = 2x + 6
　　y = -3x + 1

(b) y = -.5x – 2.5
　　y = x – 7

(c) y = 5
　　y = 15x – 7

7) How do your estimates in question 5) compare with your algebraic solutions in question 6)?

8) If 1 gallon is the same as 4 quarts, and 4 quarts is the same as 128 ounces, then 1 gallon is the same as _____

9) Find the mistake in the following solution to the system and re-work it correctly.

$y = x + 2$
$4y = -5x + 10$

$x + 2 = -5x + 10$
$\underline{+5x \qquad +5x}$

$6x + 2 = 10$
$\underline{\quad -2 \quad -2}$

$6x = 8$
$x = 8/6 = 4/3$

$y = 4/3 + 2 = 10/3$

<u>Solution</u> (4/3, 10/3)

10) List the 3 types of linear systems and the number of solutions in each system.

11) f(x) and g(x) form a system of linear equations. $f(x) = 4x - 7$.

    (a) Write 2 examples of g(x) so the system has no solutions.

    (b) Write 1 example of g(x) so the system has an infinite number of solutions.

    (c) Write 2 examples of g(x) so the system has one solution.

12) Solve the following systems using the method of substitution.

    (a) $4y = x - 16$
         $y = \frac{3}{4}x - 2$

    (b) $6x - 2y = 18$
         $3x + 5y = 27$

    (c) $y = .5x - .25$
         $1.5x - 3y = .75$

    (d) $5y - 10 = 20x$
         $4x = -6 + y$

13) Celsius as a function of Fahrenheit is given by $C(F) = 5/9\ (F - 32)$. Fahrenheit as a function of Celsius is given by $F(C) = 9/5\ C + 32$. At what temperature do these two scales meet?

14) You have exactly $537 to spend on party gifts for your rich uncle's birthday party. You decide to get watches for the ladies (at $27.98 each), and beepers for the men (at $23.46 each). You know that the number of watches required will be 3 times as much as the number of beepers. How many of each item do you buy?

15) Electro-fishing is a process where fish are stunned by introducing electricity into the water. The fish, although unharmed, are very still for taking length measurements, etc. Roy Bixby decides to do some electro-fishing for largemouth bass and perch. His total number of fish caught for the day was 18. When he added up the total lengths of all the fish, he got 239 inches. If we assume that largemouth bass are always 16 inches, and perch are always 9 inches, how many of each fish did he net?

16) Given the following diagram with point P (4,6) on line M

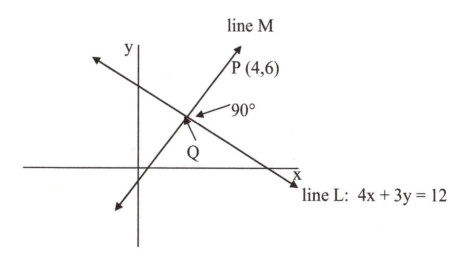

line M

P (4,6)

90°

Q

line L: 4x + 3y = 12

(a) Find the equation of Line M

(b) Find the coordinates of point Q

(c) Find the distance between Q and P (4, 6)

17) The sum of two numbers is 150. The difference of three times the second number and two times the first is 165. Find the 2 numbers.

18) Zach and Emily have competing latte stands in the village. They both have their own homemade recipes and they both happen to have their own daily profit functions for their stands. (Remember, profit is revenue minus starting costs). Zach's daily profit function is $Z(x) = .75x - 10$, and Emily's is $E(x) = 1.25x - 14$, where x is the number of latte's they sell on any given day, and $Z(x)$ and $E(x)$ represent how much profit they make per day. How many latte's do they need to sell to have the same profit?

19) X is Xavier's age, Y is Yvette's age, and Z is Zevon's age. Yvette is twice as old as Xavier. If Zevon's age is increased by 10 he will be as old as Yvette. Write an equation that shows Zevon's age (Z) only in terms of Xavier's age (X).

20) Use the Intersect method on the TI-83 to go back and check your solutions to problems 13, 14, 15, 17, and 18. If the TI gives you different answers, go back and check your work.

21) Solve the following 3 x 3 systems of equations. Be sure to check your solution in all 3 original equations.

(a) $x+y+z=1$
 $3x - y - 2z = 0$
 $x = y$

(b) $y = x + z + 6$
 $2z - 4y = 7 + x$
 $z - 5 = 3y$

(c) $\dfrac{r}{3} + 2s = 6t$

 $\dfrac{2s}{5} - t + r = 0$

 $\dfrac{t}{2} + \dfrac{2s}{3} = 4$

(d) $A + B + C = 1$
 $B - C - A = 0$
 $2C + 3A = 1 + B$

22) The sum of three consecutive odd integers is 75. The difference of the largest and the smallest is 4. Twice the middle integer is the sum of the largest and smallest. Find the 3 integers.

Extended and *Thinking* Questions

Proof that $0.\overline{9} = 1$

1) Prove to yourself that $.\overline{9} = 1$. What I'm suggesting is that .999999… (repeating indefinitely) is *exactly* equal to 1.

Here's how… we start by saying that $.\overline{9} = N$ (some number). This is an equation, right?

Now create a second equation by multiplying both sides of equation 1 by 10. Write your answer.

Next, subtract equation 1 from equation 2 and write the resulting equation below…

Divide both sides by the coefficient of N. Do the work below…

You end up with N = _____. However, we originally said that N = $.\overline{9}$. So what does that say?

The Matrix Method

A *matrix* (unlike the one that Keanu Reeves lived in) is simply a grouping of numbers in square brackets. Below are some examples of matrices (note the plural is *matrices*, not matrixes)…

$$\begin{bmatrix} 1 & 0 & 0 \\ 0 & 1 & 0 \\ 0 & 0 & 1 \end{bmatrix} \quad \begin{bmatrix} 1 & 2 & 3 \\ 6 & 7 & 9 \end{bmatrix} \quad \begin{bmatrix} 0 & -1 \\ 5 & 4 \end{bmatrix} \quad \begin{bmatrix} 11 & 9 & 4 & 6 & 9 \\ 17 & 26 & 24 & 14 & 15 \\ 15 & 15 & 15 & 27 & 27 \\ 4 & 5 & 2 & 16 & 9 \\ 21 & 16 & 26 & 12 & 5 \end{bmatrix}$$

Without getting into all of the matrix algebra theory, I would like to show you how to solve systems of linear equations with matrices on the TI-83…

If you consider any two linear equations, you will notice that the only differences are the numbers (or coefficients) of the variables x and y.

For example, $3x + 5y = 10$ and $-6x - y = 7$ are different lines only because the numbers are different.

So, it is the coefficients (not the variables) that are the "meat" of the equations. A matrix is simply a grouping of these numbers to solve linear equations. We get rid of the variables (x and y in this case) and just deal with the numbers.

We can put the 2 x 2 system mentioned above into a matrix that looks like this…

$$\begin{bmatrix} 3 & 5 & 10 \\ -6 & -1 & 7 \end{bmatrix}$$

Notice we have the coefficients of x in column 1, coefficients of y in column 2, and the constants in column 3. In general, it looks like this…

$$\begin{bmatrix} x_1 & y_1 & c_1 \\ x_2 & y_2 & c_2 \end{bmatrix}$$

We then use a command called reduced row echelon form (or rref) on this matrix, and we get…

$$\begin{bmatrix} 1 & 0 & \frac{-5}{3} \\ 0 & 1 & 3 \end{bmatrix}$$

The way we read this matrix is $1x + 0y = -5/3$….in other words $x = -5/3$ in row 1, and $0x + 1y = 3$…. or $y = 3$ in row 2. Our solution is (-5/3,3).

Using the TI-83 Matrix commands

We will solve the same system using the TI-83.

Step 1: Press the MATRX button, which is 2nd , x^{-1} and you will see this window…

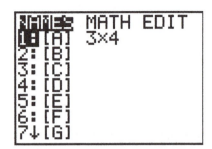

You may or may not have any numbers in the MATH column. The 3x4 means that matrix A is a 3 row by 4 column matrix. Our matrix A (from above) has 2 rows and 3 columns, so it is a 2x3 matrix. We read it as "two by three matrix."

Step 2: Arrow over to EDIT and press ENTER, and you will see this…

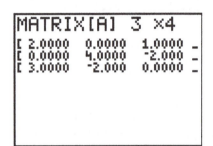

On your TI-83, the cursor on the 3 should be blinking…

Don't worry about the numbers below because we are going to change them anyway.

Step 3: Type "2" over the blinking cursor, hit the RIGHT ARROW and type "3". You will see

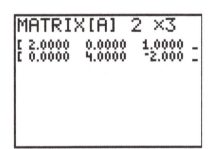

Step 4: Now we want to type in our matrix entries from the system above…
$$\begin{bmatrix} 3 & 5 & 10 \\ -6 & -1 & 7 \end{bmatrix}$$

Type "3" and then ENTER, "5" and then ENTER, and so on until your TI-83 matrix looks like…

```
MATRIX[A]  2 ×3
[ 3.0000   5.0000   10.000 ]
[ -6.000  -1.000    7.00000 ]




2,3=7
```

Step 5: Press 2nd MODE and go to the home screen. Clear the home screen.

Step 6: Press **2nd** , **x^{-1}** and RIGHT ARROW to MATH. You will see this window…

ARROW DOWN to **B: rref(** and press ENTER. You will see the following on the home screen…

rref (

Step 7: Press **2ⁿᵈ** , **x⁻¹** and **1** and then close the parentheses **)**

You should see **rref([A])** on the home screen

Step 8: Press Enter and you will see…

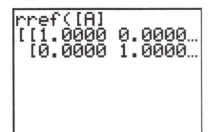

Use the RIGHT ARROW key to scroll over to the final entries and you should see   -1.6667]
                                                                                                              3.0000]

Those are the **answers**…x = -5/3 and y = 3.

Exercise using The TI-83 matrix method.

Your campus opens up a new grease-pit of a restaurant called "Algebrelli's". The meal names, number of items, and total costs are summarized in the table below.

Determine the <u>individual price</u> of each of the five food items by setting up the equations and using the Matrix Method on the TI-83. Hint: you will be setting up a 5 x 6 matrix The empty boxes must be represented by 0's…in other words 0x, 0y, etc.

Name of Meal	Cheese Burger	Slice of Pizza	Medium Fries	Side Salad	Choice of Drink	Meal Cost
The Overweight Special			1		1	$1.75
The Meat & Potato	2		2		3	$8.95
The Neapolitan		2		1	1	$5.80
The Hungry Man	4		1		1	$10.75
The Yikes!	3	1	2	1	3	$14.30

Booklet on Exponential Functions

*T*he Exponential Family of Functions

Is it possible to hear $\sqrt{2}$ *?*

In the Booklet on Basics of the Language, we looked at the Western Chromatic Scale and used it to "hear" $\sqrt{2}$ and to help develop rational and irrational numbers. We showed that in order to find the frequency of the next note in the scale we multiplied the preceding note by $2^{1/12}$.

From that fact, we generated the following table...

Note	Frequency (Hz)
C	1000.00
C#	1059.46
D	1122.46
D#	1189.21
E	1259.92
F	1334.84
F#	1414.21
G	1498.31
G#	1587.40
A	1681.79
A#	1781.89
B	1887.75
C	2000.00

CALCULATOR EXERCISE
(time to practice pressing buttons!)

As you have done already a *cajillion* times, enter the data above into 2 lists on your TI. Since we cannot enter actual *notes* for x values, change the first column to the numbers 1 through 13. ((How many 0's are in a cajillion anyway?))

Enter the 2nd column as is. Adjust your viewing window to be 0, 15, 1, 0, 2100, 1, 1.

What you should see on your TI after pressing graph is the following:

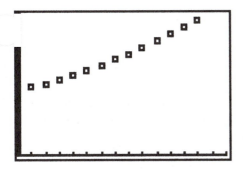

Find the exponential regression equation (the exponential equation that best fits the data) using the **ExpReg** option from the STAT and CALC menus. Before you do this, though, change the number of decimal places on your TI to **5**, by pressing MODE, arrow down to FLOAT and over to 5…press ENTER.

Run the regression and write the following from your TI screen.

ExpReg
$y = a*b^\wedge x$
a = _____
b = _____
r = _____

NOTE: If you do not see the "r = number" line on your screen under a = and b =, go back to the <u>Booklet on the Linear Family</u> to revisit the TI steps necessary to make the r value appear.

How can check to see if your equation is a good match for the data?

Type the equation above in **Y=** (or do it the quick way!...what! you don't remember the quick way to do that…???)

Once the exponential regression equation is in **Y=**, press **GRAPH**. You should notice the graph (from the equation) *exactly* match the scatter plot (from the data). If you remember our lessons from regression analysis, you probably already knew this was going to happen because the r-value is 1. <u>An r-value of 1 means the data exactly matches the equation.</u> The previously underlined sentence will probably be on your next quiz ☺..Psst!! Don't tell your professor I said that – he's probably not reading this anyway!

At this point you probably have some more questions…

...Why does the shape of the data look *curved*?
...What do the values for *a* and *b* mean?
...Why did we use exponential regression instead of *linear* regression?

These are all good questions. And we will answer them shortly. But, before we do, we should probably take a look at what an *exponential function* really is...what does it look like? What are its properties and characteristics? Where does it live?..etc...

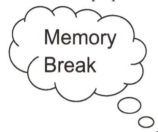

Memory Break

Brief Review of Exponents

Exponents are the little numbers (or letters) on top of the big numbers (or letters). An exponent tells you how many times to multiply the thing under it by itself. For example, x^3 means multiply x by itself 3 times, or (x)(x)(x). Another way of saying that is "x raised to the 3rd power, or x-cubed." It works the same way with numbers...$4^5 = (4)(4)(4)(4)(4) = 1024$. Notice that 2^{10} is also 1024. We halved the <u>base</u>, so we double the <u>exponent</u> to get the same quantity.

What if we replace the base of an exponential expression with a variable?

$$5^2 \text{ becomes } x^2$$

If we do this, we change a number (25) to a function. In this case, a *quadratic* function – a family that is dealt with in more detail in another Booklet.

What if we now *transpose* the base and the exponent?

Warning – big word alert

$$x^2 \text{ becomes } 2^x$$

Now the base is a number and the exponent is the variable. Functions of this form are called *exponential* functions.

Examples of exponential functions...

$f(x) = 3^x$

$g(s) = 10^{s+1}$

$h(x) = e^x$

$y = 5 \cdot 2^x$

$T(r) = 1.407 \cdot 5.38^r$

$p(x) = 10^x$

The examples above are all different, yet somehow similar. Remember we said this was a characteristic of families of functions. All of the equations above have 3 characteristics in common: a coefficient, a base, and a (variable) exponent. ((The coefficients of f(x), p(x), g(s), and h(x) are all 1))

Any variable is just as good as any other variable, so let's use the TI variables to limit confusion.

Properties of the exponential family of functions

A member of the exponential family of functions has the equation…

$$f(x) = a \cdot b^x$$

<u>a</u> is the coefficient and can be any real number except 0…if it's 0, there's no function at all, right?

<u>b</u> is the base of the function and must be positive (> 0) and also not equal to 1 ($\neq 1$).

<u>x</u> is the (variable) exponent and can be anything. Remember x is *not* just x. x can be anything: x + 1, 3x, 10x – 4, 0.5x, etc…variables are simply shoeboxes.

Why does b need to be positive and not equal to 1 (b > 0, b ≠ 1)?

- If b = 1, we would have 1^x, and we know that 1 raised to anything is 1, so the function would be f(x) = a·(1), which is just f(x) = a…some number. So f(x) would not be an exponential function at all, but a horizontal line. And, as *SnowMeiser* likes to say on *The Year Without a Santa Clause*, "…who needs that??"

- Also, If b = 0, the entire function is 0…that's boring!

- The tricky part is what happens when b is *negative*…Let a = 1, b = -2, and x = 1/2, so $f(x) = (-2)^{1/2}$. Another way to write that is f(x) = √-2, which we know is not a real number. In order to keep our functions *real*, b must be greater than 0.

What does a member of this family typically *look* like…?

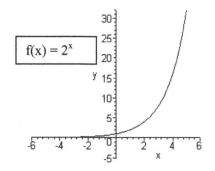

Let's point out several features:

1) The curve rises fairly quickly.

2) The curve does not go below the x-axis, but gets closer and closer to the x-axis in the negative x direction.

3) The curve has a y-intercept.

What do other family members look like?

The questions below the graphs are not rhetorical. Think about the answers and write them out, write questions for your professor, write the local pizza delivery number…Write in the book, whenever and wherever you can. Remember, you are *not* selling it back!!

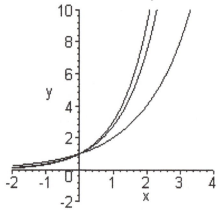

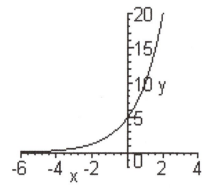

Graphed above are 2^x, 3^x, and e^x.
Which graph is which function and why?
Where do all 3 graphs cross the y-axis?

The function above is $5 \cdot 2^x$.
Where does this graph cross the y-axis?

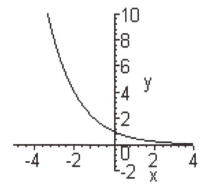

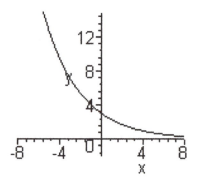

Graphed above is $(1/2)^x$.
Why do you think the graphed "flipped"?
Where is the y-intercept?

The function above is $3 \cdot (0.75)^x$.
Why did it flip again?
Where is the y-intercept?

Based on these examples and questions, what can you _deduce_ about the values of **a** and **b**, and how they affect the shape and properties of the graph?

Write stuff here…

I asked you to deduce something about the values of **a** and **b**. What does it mean to _deduce_ something? Sherlock Holmes was the master of deduction. He started with a bunch of evidence and combined and summarized it into one suspect. See the upside-down pyramid below. The opposite of deduction is _induction_. To induce something, you start with a basic premise and then make application to a wider group. See the other pyramid below…

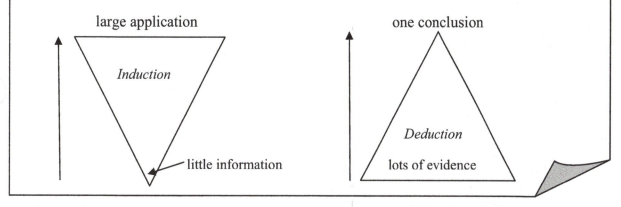

If you're still not sure, let's look at one more example…a family portrait, if you will…

f(x) = $2(0.75)^x$ $2(0.25)^x$ $2(2)^x$ $2(5)^x$

    $2(0.5)^x$ $2(3)^x$

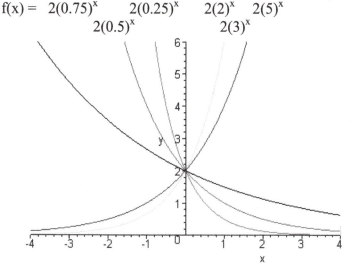

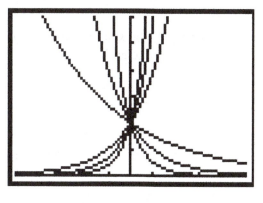

TI-83 version

Hopefully, by studying these different members of the exponential family of functions, you have been able to deduce some properties of *a* and *b*.

Property 1: The point (0, a) is the y-intercept. Why? $f(0) = a \cdot b^0 = a(1) = a$. (input, output) = (0, a)

Property 2: The value of b determines whether the curve is increasing or decreasing. If b > 1, the function increases. If 0 < b < 1, the function decreases. Remember we always read math like we read English…from left to right.

By the way, what happens if a < 0?

Think back to the Booklet on Functions…When f(a) becomes f(-a), what happens? In the box below, sketch $f(x) = -(2)^x$. Note that $-(2)^x$ is not the same as $(-2)^x$, which is not allowed.

Property 3: The Domain of all members of the family of exponential functions (with a > 0) is *all real numbers*. Another way to express that is $-\infty < x < \infty$. In other words, there is no leftmost or rightmost x-value.

Your sketch of $f(x) = -(2)^x$

Property 4: The Range of all members of the family of exponential functions (with a > 0) is *all real numbers greater than 0*. Another way to express the Range is y > 0. Look at the colored graphs. 0 is not included in the range because the graphs never meet the x-axis.

How are you doing on that sketch over there?? Did you finish?
If you did, check your answer on the TI by putting the function in **Y=** and pressing **ZOOM 6**.

Property 5: The x-axis is an *Asymptote* for the family $f(x) = a \cdot b^x$. An Asymptote is a line that a function approaches but never touches…it just keeps getting closer and closer.

Math to English

The math symbol → means "approaches"

Given the function $f(x) = 2^x$, for example, we say that as $x \to \infty$, $f(x) \to \infty$. Also, as $x \to -\infty$, $f(x) \to 0$. In English, as x gets larger and larger, the graph also gets larger and larger. As x gets larger in the negative direction, the graph gets closer and closer to 0. Take another look at 2^x above just to be sure.

Now we can go back to the WESTERN CHROMATIC SCALE. Your calculator screen for the ExpReg should have looked like this…What I'm saying is these are the correct answers!

What do the answers mean??

The exponential equation that best fits the Western Chromatic scale (tuned to C, 1000 Hz) is

$$f(x) = 943.85824 \, (1.05947)^x$$

The y-intercept is the point (0, 943.9) when rounded.
The function increases because $1.05947 > 1$
The equation *exactly* matches the data because $r = 1$
 (that is, all points fall on the graph)

ExpReg
$y = a*b\wedge x$
$a = 943.85824$
$b = 1.05947$
$..r = 1.00000$

Notice that $f(1) = 943.85824 \, (1.05947)^1 = 1000$ Hz (the frequency of the note C).

What in the world is 1.05947? On your TI-83, evaluate $2^{1/12}$. Ahh Haaah!!

So, we could write our function as $f(x) = 943.9 \cdot (2^{1/12})^x$.

Professor's Practice Pause

1) In reality, the accepted tuning of C is actually 1046.5023Hz and A is 440Hz. This is considered concert tuning. If we used concert tuning instead of 1000Hz, write the new function for the Western Chromatic Scale.

2) Describe (in writing) all of the graphical characteristics of the function, $f(x) = 2.5 \cdot 0.86^x$

3) An exponential function is increasing and crosses the y-axis at (0, 14). If every successive output is three times as large as the previous one, write a possible equation for this function.

Associated End of Booklet exercises are 1 – 24, 49, 52, and 53

Population growth and decay

Besides music and money (which we will study later), there are many other real-life variables and situations that can be modeled by exponential functions; probably more so than any other family. The decay of radioactive elements is governed by exponential functions (you've probably heard of *half-life*). Just about every instance of population growth (or decay), whether it be human, animal, bacteriological, viral, or even artificially – intelligent programming can be represented by some function in the exponential family. That's why the author (that's me) considers this family (and the logarithmic family) to be more important to real-life situations than the other families of functions.

Example: The time it takes for a bacterial cell to divide using binary fission, or a population of cells to double is called the *generation time*. The Center for Disease Control (CDC) lists the *Escherichia coliform* bacteria (or E. coli) as having a generation time of 12 – 24 hours in a person's intestinal tract. Let's just assume 24 hours for minimal sickness…Uhmm…I mean… minimal *growth*. The table below compares the number of days versus the number of bacterial cells.

Number of days of infection (d)	Number of bacterial cells (c)
0	5
1	10
2	20
3	40
4	80
5	160
10	5120
14	81920
21	1.05×10^7

Based on the data, the number of cells is clearly a function of the number of days. (Why? If you're not sure, go back to the Booklet on Functions and take another look at the definition). We can write $c = f(d)$. Since $f(0) = 5$, we know that the point (0, 5) is the y-intercept. That means for an equation so far we have… $f(d) = 5 \cdot b^d$, but what is the base? How do we figure that out??

Growth (and Decay) Factor

The growth (or decay) factor of an exponential function is just another name for the base. We can find this factor (b) by looking at *successive ratios of outputs*. Whoa!! Big phrase right there…Successive means in a row, ratio means fraction or divide, and the outputs are the y values (or output values, or graph values, or functional values). So, in the table above, we would make the following divisions to see if they are all equal…

10/5, 20/10, 40/20, 80/40, etc... They are all equal and all equal to 2. So, the base of the function is 2. We call this the **Doubling Function** since successive outputs double. The function 3^x would be the Tripling Function, and so on.

Very Important Point: Notice that 5120/160 is *not* 2. Why? Because those outputs are not successive (not in a row)...the inputs jump from 5 to 10 (not 5 to 6).

How do we know whether we have growth or decay? If the base equals 1 (which, remember is not allowed), we would have no growth or decay, just a horizontal line. So, the number 1 is the dividing line. Every base greater than 1 grows and every base less than 1 will decay.

CALCULATOR EXERCISE

Go to **Y=** and type in .75^x for **Y₁** and 1.25^x for **Y₂**. You should see the graphs below...

Y₁=.75^x is decreasing (decaying) from right to left.	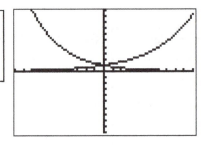	**Y₂**=1.25^x is increasing (growing) from right to left.

Professor's Practice Pause

1. Using the days of infection versus number of bacterial cells data, answer the following:

    (a) Write the function for number of cells being dependent upon the number of days.

        $f(d) = $ _____

    (b) Calculate f(7) and explain what information that gives you.

2. What does the following math statement "$f(28) = 1.34 \times 10^9$" mean in English?

3. Antibiotics kill bacteria. A person with E. coli who begins taking antibiotics will turn the exponential growth into exponential decay. Which of the following functions could *possibly* represent that decay? Why?

    1. $f(x) = 200,000(3^x)$
    2. $f(x) = 200,000(.8^x)$
    3. $f(x) = 200,000(1.4^x)$

4. Write a function that describes bacterial growth (g) as a function of minutes (m). The bacteria population starts at 1500 and *quadruples* every minute. Let the function name be G.

Associated End of Booklet exercises are 34 – 39, 54

On the Waterborne Pathogen Information Sheet, distributed by the United States Department of Agriculture (USDA), the particular strain of E. coli that causes troubles for humans is O157:H7. As of the year 2000, the USDA estimated about 20,000 cases of E. coli O157:H7 each year in the United States, 250 of which were lethal. As little as 10 cells are needed to start an infection.

It should also be noted that exponential growth is only part of the natural life-cycle of bacteria. Out "in the wild" (as opposed to the laboratory), bacteria colonies start growth very slowly and only then enter the exponential growth stage. The growth eventually hits a plateau, stays there for a bit, and then they begin to slowly die. The growth curve looks more like a flattened *Bell Curve* (Normal Distribution from statistics).

Warning – big word alert

Of course this growth cycle can be immediately interrupted with the use of Lysol® brand disinfectant, which kills 99.9% of germs in 30 seconds (at least that's what it says on my can). Fortunately, *E. coli* is one of those germs! My favorite scent is *Purple Putrescence.*

Growth (and Decay) Rate

The growth or decay of an exponential function can often be expressed as a rate (instead of a factor). But, luckily for you, they are connected. If we write the base (b) as $1 + r$ or $1 - r$, depending on whether or not b is greater than or less than 1, the r is the *rate*!

HELPFUL EQUATION $b = 1 \pm r$

Example: Start with the function $H(z) = 8(.63)^z$. Let's do this question and answer style, kind of like a *Reader's Digest* interview with a celebrity…or then again, maybe not!

Q: What is the initial value (y-intercept) of the function?
A: $H(0) = 8(.63)^0 = 8(1) = 8$… the value of *a*. The y-intercept is always (0, *a*). In this case (0, 8)

Q: Is this function growing or decaying?
A: Because the base is .63, which is less than 1, the function decays.

Q: What is the decay factor?
A: The decay factor is the same as the base, .63.

Q: What is the decay rate?
A: Since the base is less than 1, we can write $b = 1 - r$, or $.63 = 1 - r$. Since $1 - .37 = .63$, the decay rate is .37, or 37%. That means for each increase of one input value (z), the output values, H(z), will decrease by 37%. This function would be good against bacteria, but not too good as a bank account! Note that multiplying by 63% is the same is decreasing by 37%.

$H(0) = 8$
$H(1) = 8(.63)^1 = 5.04$. Notice that 37% of 8 is 2.96. $8 - 2.96$ is 5.04.
$H(2) = 8(.63)^2 = 3.175$. Notice that 37% of 5.04 is 1.865. $5.04 - 1.865$ is 3.175.
$H(3) = 8(.63)^3 = 2$. Notice that 37% of 3.175 is 1.175. $3.175 - 1.175$ is 2.

Example: Given the function $f(x) = 5(1.48)^x$.

Q: Is this function growing or decaying?
A: Because the base is 1.48, which is greater than 1, the function grows.

Q: What is the growth factor and growth rate?
A: The growth factor is the base, 1.48. The growth rate can be written as $b = 1 + r$, or $1.48 = 1 + r$. The growth rate is .48 or 48%.

Example: What is the growth rate of the doubling function, 2^x?

A: $2 = 1 + r$, so $r = 1$ (or 100%). The function 2^x doubles because each successive output is increased by 100%.

Tricksy note, my precious!!! Note that 100% is doubling, not 200%. To double your salary, it would have to be increased by 100%. In English, "Original plus one-hundred percent of original is double the original value." If you increase something by 200%, you have actually *tripled* the original, and so on.

Real - World Growth

Example: California is the most populated state in the Union. The US Census Bureau gives the following estimates for total population of California from 2000 to 2007.

Year	Population (in millions)
2000	33.871
2001	34.600
2002	35.116
2003	35.484
2004	35.893
2005	36.132
2006	36.457
2007	36.553

Is California experiencing exponential growth?

Calculate the following quotients of successive outputs:

34.6/33.871 = _____

35.116/34.6 = _____

35.484/35.116 = _____

36.457/36.132 = _____

36.533/36.457 = _____

Are the quotients all the same? _____

What would that lead you to believe?

What pattern do you see in your answers (going from top to bottom)?

What does that tell you about the population of CA from 2000 to 2007?

Maybe the growth is linear instead!?! Pick any 3 pair of years and calculate the slope between those points…

Pair 1: Year_____ & Year _____

m = Change in Outputs / Change in Inputs =

Pair 2: Year_____ & Year _____

m = Change in Outputs / Change in Inputs =

Pair 3: Year_____ & Year _____

m = Change in Outputs / Change in Inputs =

Q: Are your slopes the same?_____

Q: What does that lead you to believe about the *linearity* of this function?

Using your best British accent, say the following:
"I believe we've run into a bit of a quandary! You see, the data neither support an exponential function, nor do they support a linear function. Quite!"

QUICK CONCEPT SUMMARY:

1. To tell if data is <u>exponential,</u> we check to see if *quotients of* successive outputs are equal.

2. To tell if data is <u>linear,</u> we check to see if *differences between* successive outputs are equal.

Back to the WESTERN CHROMATIC SCALE yet again…This represents another illustration of the difference between linear and exponential functions, but also shows an example of both happening at the same time. In the Booklet on the Linear Family of functions, we saw that the progression from one note to the next in an octave could be described by a linear function. Furthermore, and at the same time, we see that the increase in frequency as you progress from note to note is exponential. That's awesome!

The note (or step) in the octave is a function of number of keys (distance). This function is <u>linear</u> since the step level changes by a constant value ($m = \frac{1}{2}$) per key.

The frequency of the note is also a function of number of keys (distance). This function is <u>exponential</u> since each successive note increases the frequency by a factor of $2^{1/12}$.

STUDENT EXAMPLE TIME: Can you think of a situation with several variables where two variables act linearly, but another two variables act exponentially (or anything other than linearly)?

_____

_____

I hate to do this to you…but we will revisit this problem when we get to the section entitled *"Modeling Exponential Functions using the TI-83."* *Quite!*

Real - World Decay

The half-life of a radioactive element is the time necessary for $\frac{1}{2}$ of the atoms to disintegrate into a more stable form (become safe, so to speak). Half-lives can range from a few seconds to billions of years. This time is also unique to each radioactive element, which has led to the very popular phrase, "You can tell a nuke by its half-life!" Of course this phrase is no longer repeated publicly, except maybe in the mining bars of West Virginia…Anywho…

Radioactive elements undergo exponential decay, given by the following function:

$$A(t) = A_0 e^{-ct}$$

Where,
A(t) is the amount of material present after time t
     A_0 is the initial amount of material
     e is our friendly neighborhood irrational constant
     t is time (usually in years)
     c is a decay constant, unique to each element.

Just as an example of the different half-lives, to show you that I'm not joking, here are three half-lives for different isotopes of Uranium:

U-232	72 years
U-237	6.75 days
U-240	14.1 hours

Example: For U-232, the decay constant (c) is approximately .00963. ((I will teach you how to solve for the decay constants when we get to the *Booklet on Logarithmic Functions*)).

Q: Given an initial quantity of 500g of U-232, how much will remain after 216 years?

A: $A(216) = 500e^{-(.00963 \cdot 216)} = 500e^{-(2.08)} \sim 62.5g$

Notice something very peculiar…62.5 is one-eighth of 500. In other words, $500(1/2)(1/2)(1/2) = 62.5$. Also, the years follow the same pattern…

After this many years…	The amount of material is…
72	250…(½ of 500)
144 (72 + 72)	125…(½ of 250)
216 (72 + 72 + 72)	62.5…(½ of 125)

This tells us that the half-life is a *constant* for every radioactive isotope. For U-232, every 72 years reduces the amount of material by ½. For U-237, every 6.75 days reduces the amount of material by ½, and so on…

Investments and interest (a second derivation of e)

As I write these words, the New York Stock Exchange (NYSE) has gone from over 13,000 down to 11,000, and back up again in the space of a week. Nothing seems to be more susceptible to changes in the world than the stock market. I find it amazing that even rumors of something happening are enough to bring about drastic changes one way or the other.

I would like to introduce you to some basic Investment concepts. Hopefully, you will learn something about your own investments, or learn how to invest if you currently have none, and maybe make some changes to help you earn more money!! Sweet! And, I'm not even going to charge you for that – it comes with the book!

Back in the Booklet on Basics, we studied our Aunt Sally...well not *exactly* our Aunt – we looked at order of operations and *remembered* that with the acronym PEMDAS (Please Excuse My Dear Aunt Sally...or whatever acronym you made up!). One of the expressions we evaluated was used for calculating *simple interest*...$P(1 + i)^t$. If that's fading from your memory – which is possible since it was at least a month ago – go back and refresh.

Simple interest means that the bank or financial institution calculates your interest once per year and adds the interest to the principal at that time. The process of the bank doing that is called *compounding*. So, with simple interest, the bank compounds once per year.

The standard for today's banks is to compound more than once per year. If you have a savings account or CD (Certificate of Deposit) with a bank you should find out 2 things: (a) What your interest rate is, and (b) How often the bank compounds your interest. We use the letter *n* to represent number of compounding per year, which allows us to create the following table:

Portion of year	number of compounding (n)
Annually (simple interest)	1
Semi-annually	2
Quarterly	4
Monthly	12
Weekly	52
Daily	365

...you get the idea.

The compound interest formula is as follows...

$$A = P\left(1 + \frac{r}{n}\right)^{(nt)}$$

A is amount of money, **P** is principal (initial investment),
r is interest rate (in decimal), **n** is the number of compoundings per year,
t is the number of years.

EXAMPLE: Your distant relative leaves you $10,000 in her will. You decide to invest the money in a CD at 4.5% compounded quarterly for 7 years. Assuming you neither deposit nor withdraw any money, how much will you have in the account after 7 years?

Before simply substituting values let's first assign the values to the appropriate variable...

A is what we are solving for, P = 10000, r = 4.5/100 = 0.045, n = 4 (quarterly), and t = 7

Using $A = P\left(1 + \dfrac{r}{n}\right)^{(nt)}$ we have A = **10000(1 + .045/4)^(4*7)** = 10000(1.01125)^28 = $13,678.52. Which means we made $3,678.52 in interest over the 7 years without doing a thing...not bad!

Notice the way I typed the **bolded expression** above is <u>exactly</u> how you would enter it in the TI.

```
10000(1+.045/4)^
(4*7)
          13678.516
```

Another method of compounding that banks use today is **continuous compounding** (although it is typically called the *Annual Yield*). Weekly means every week, daily means every day, etc. But what if you compounded every minute, no... every second, no... every 10th of a second, or 100th of a second, or microsecond...this is compounding continuously. The interest is *always* compounding. Because of that compounding continuously represents the most amount of money you can make with a fixed principal, rate, and time.

So how do we do this mathematically? We can't simply let n be infinity...that wouldn't make sense.

$$A = P\left(1 + \frac{r}{\infty}\right)^{(\infty t)} \quad \text{VERY SILLY!!!}$$

Let's say P = 1, t = 1, and r = 1. That means our equation looks like A = (1 + 1/n)^n. What happens as n gets bigger and bigger?

CALCULATOR EXERCISE
(once again...time to practice pressing buttons!)

Enter the expression above in **Y=**. Since we cannot use n, we'll switch to x...**Y**= (1 + 1/x)^x
Now go to **TBLSET** and let `Indpnt` be `Ask`

Open up your table and enter the numbers for x as below and fill in your outcomes for Y1...

X	Y1
2	
5	
10	
100	
10,000	
1,000,000	

As n approaches infinity (gets bigger and bigger) the factor of $(1 + 1/n)^n$ approaches a fixed number ~ 2.718. And there is the number e once again. Remember from the Booklet on Real Numbers, e is an irrational constant that we derived from factorials.

Back to compounding continuously... We can replace the $(1 + 1/n)^n$ with e^1. If you do the same TI exercise with $(1 + 2/n)^n$ you will end up with the values approaching 7.389, which is e^2.

$(1 + 3/n)^n$ will approach e^3, and so on.

this part is e^r

So, $(1 + r/n)^n$ will approach e^r, and so the equation $P(1 + r/n)^n$ becomes $\mathbf{Pe^{rt}}$. This is the equation used for compounding continuously.

Which bank account yields the most money?

We now revisit your poor, deceased, distant relative who left you $10,000 in her will. You decide to invest, invest, invest!!! Since you are getting married in **5 years**, a little nest-egg would be a great thing to start out with, eh? You traipse into 3 local banks (la, la-la) to shop around a little...but how do you know which account will earn you the most money? They're all different. Your mind races with the possibilities...

"Should I go for more compounding and lower interest rates or maybe less compounding with higher interest rates?" AGHHH! Too much information!!

Let me help you...below are the three accounts:
World of Wealth will compound interest quarterly and give you 5.25%
Continuous Credit will compound interest continuously and give you 5.05%
Looey's Local Loans will compound interest quarterly and give you 5.05%

World of Wealth **(WW)**
$$A = 10000(1 + 0.0525/4)^{(4 \cdot 5)} = \$12,979.58$$

Continuous Credit **(CC)**
$$A = 10000 \, e^{(.0505 \cdot 5)} = \$12,872.39$$

Looey's Local Loans **(LLL)**

$$A = 10000(1 + 0.0505/4)^{(4 \cdot 5)} = \$12,852.06$$

Compare CC and LLL...they have the same interest rate, but a different amount of compounding. CC has the higher compounding so we would expect to make more $$ there...and we did -- $20.33...big deal! However, it is true that more compounding gives more money earned.

Translation: Lots of money!

Compare WW and LLL...they have the same amount of compounding, but different interest rates. WW gives the higher rate so we would expect to make some serious fat squirrel...and we do -- $127.52. Big deal! However, it is true that higher interest rates yields more money earned.

The question is which is better? Do you go with the higher interest rate or more compounding?

Let's see if you can answer that question by considering my very own graphic below. The illustrator did, in fact, draw this graphic, but it is <u>my</u> very own graphic!

n means $ ...but   r means $

Increasing the number of compounds will earn you more money, but getting a higher interest rate will earn you *even* more money. So, if all things are equal in different accounts, go for the higher interest rate...even if it is only one-tenth of a percent higher. You will still reap higher profits. Psst!! That question will probably be on one of your professor's quizzes. *Which bank account will give the highest return? The one with the highest interest rate* (even if n is lower).

Professor's Practice Pause

Use your TI to find the amount of money earned in the following scenarios. Write out the expression in the space provided and then type the entire expression in one line, as we did above.

(a) $5500 invested for 4 years at 3.7% interest compounded monthly. A = _____

(b) $7500 invested for 10 years at 2.7% interest compounded quarterly. A =_____

(c) $1200 invested for 5 years at 4.25% interest compounded continuously. A = _____

(d) Without doing any math, explain which of the following accounts will earn you more money:
      1. 3.5% compounded quarterly
      2. 3.6% compounded biannually
      3. 3.4% compounded continuously

Associated End of Booklet exercises are 25 – 33, 40 – 48, 50, 51, 55 – 57

Modeling Exponential Functions using the TI-83

For exponential regression on the TI-83 we want to go to **STAT**, **CALC** and then option Ø:ExpReg. Everything else is exactly the same as when we learned linear regression. So, back to the California population problem...

The Data Trick... Instead of starting at year 2000, start at year 0 when you enter the data into the TI-83 list. Use 0, 1, 2, etc. for the years

Year	Population (in millions)
2000	33.871
2001	34.600
2002	35.116
2003	35.484
2004	35.893
2005	36.132
2006	36.457
2007	36.553

Based on the work we did a few pages ago, we know that this data doesn't exactly fit a member of the exponential family, nor does it fit (exactly) a member of the linear family. Is one better than the other? What does it fit? Why are we doing this? Stop asking questions! Oh, sorry – that's me asking the questions, isn't it?

Enter the data in **L1** and **L2** on your TI-83, and find the line of best fit (LinReg) and the exponential of best fit (ExpReg). Write the two equations and their corresponding r-values below. ((Remember if the r-values do not show up, you have to press **2nd 0**, scroll down to DiagnosticOn, and press **ENTER** twice. The TI should show Done. Run your regression again and you should see the r-values)).

Equation for the line of best fit: _____   r-value: _____

Equation for the exponential of best fit: _____   r-value: _____

Based on the r-values, which model **best** fits the data? _____

How do you know that's true? _____

```
ExpReg
 y=a*b^x
 a=34.207
 b=1.011
 r²=.951
 r=.975
■
```

```
LinReg
 y=ax+b
 a=.375
 b=34.200
 r²=.955
 r=.977
■
```

What does this example teach us about the nature of real-life data?

First, it doesn't always fit the theoretical constructs. That is, it doesn't always work out the way you think it should. We know that most population variables are best modeled by exponential functions, however in our California example, the *linear* model works slightly better. They are both extremely good since the r-values are very close to 1, but linear is just a touch better.

Second, you have to try different models. Just because the shape of the data based on a scatter plot "looks" like a certain family of functions, doesn't necessarily mean that family will provide the best model.

Third, keep in mind that this whole business of regression analysis is *very* sensitive to even small changes in data. If we were to add just one more data point, it could change the models to drastically favor one over the other, or change it so that another family altogether is better. We will do an exercise like this when we get to the Booklet on Logarithmic functions.

Making predictions...

What will the population of California be in 2010? Using both the exponential and linear models, find f(10)...

Exponential: $f(x) = 34.207(1.011)^{10} = 38.162$ million people.

Linear: $f(x) = .375(10) + 34.2 = 37.950$ million people.

Topical Summary of Exponential Functions

Take about 20 – 30 minutes and create your own summary of this chapter. Go back, review, and write below all of the main points, concepts, equations, relationships, etc. This will help lock the concepts in your brain now, and provide an excellent study guide for any assessments later.

Another featured guest in mathematical history...

Carl Friedrich Gauss

There once was a little 6 or 7 year old schoolboy named Carl Friedrich Gauss. Gauss lived back in the early 19th century...way out in the German countryside. Back then schools were pretty stern, harsh environments, and schoolmasters were even harsher. Paper was too expensive for normal, everyday folks to have, so students wrote on little slates with chalk. Communication between teacher and pupil was primarily achieved via the hickory switch.

One day the schoolmaster asked the students to add up all the numbers from 1 to 100 on their slates (i.e. $1 + 2 + 3 + 4 + ...100$). I guess it was busy work or something, I don't see any educational value in doing an exercise like that. I guess it might have been okay for the first 20 or 30 numbers...but after that it had to be time consuming. Each number was a separate calculation. Oh well ??

After about 15 seconds, Gauss puts his slate down and declares that he is finished. Of course all of his schoolmates begin to snicker - half because they think he's about to get a whoopin', and half probably because they think he's an idiot (it was Gauss' first year of school).

The schoolmaster calls him forward and is astonished to see the correct answer (5,050) on his slate with very few numbers and calculations written down. After the teacher cleared the silly look off of his face, he asked the young whip how he did it (promising a good switchin' if trickery was involved, I'm sure). I'm not exactly positive, but I believe Gauss answered something like this:

Instead of writing the numbers one at a time from the beginning, he wrote some of the first ones and some of the last ones.

$1 + 2 + 3 + 4 + 5 + ... + 95 + 96 + 97 + 98 + 99 + 100$

If you add the first and the last numbers what do you get? 101

What if you add the second and the second-to-last numbers (2 and 99)? 101

What if you add the third and the third-to-last numbers (3 and 98)? 101

See the pattern? So Gauss added 101 together 50 times (half of the total amount of numbers). Actually he just multiplied 50 times 101 and he was done?

Upon hearing his reasoning, the schoolmaster did something history thanks him for. He realized the young Gauss' mathematical prowess and set him up with a prominent, private math tutor at the local University. At 19, Gauss defended his doctoral thesis by proving the fundamental theorem of algebra in the complex plane. Every major work that Gauss has done has been the seminal work in that particular area. He is considered one of the three greatest mathematicians in history (along with Newton and Archimedes).

We can generalize this situation to be able to quickly sum-up all the numbers from 1 to n (any number). If you look at the pattern above, you are adding up one more than the highest number (or n + 1). How many times?...half of the highest number (or n/2). So the sum of the numbers from 1 to n = (n + 1) (n) / 2. I think you might remember this from your end of booklet exercises on quadratics, yes?

Try it! Add up all the numbers from 1 to 10 and see if it is the same as 11(10)/2 = 55??

The sum of all the numbers from 1 to 15456 = (15457)(15456)/2, and so on...

Dankeschoen, Herr Gauss!!

Google "Carl Friedrich Gauss" for more insight into his life and work.

End of Booklet Exercises

1) What is another way of writing 10^3? 5^4? x^2?

2) Simplify the following expression using exponents: $3 \cdot 2 \cdot 3 \cdot 10 \cdot 2 \cdot 10 \cdot 3 \cdot 2 \cdot 10 \cdot 10 \cdot 2 \cdot 3$

3) Given the equation $G(p) = 1.6 \cdot 3^p$, identify the following:

      (a) Function name   _____
      (b) Input variable   _____
      (c) y-intercept   _____
      (d) base   _____

4) Given the equation $f(x) = 5 \cdot 2^x$, *explain* how to evaluate $f(3)$.

5) In no uncertain terms, explain or list at least 3 differences between exponential and linear functions.

6) Explain what is meant by the phrase "exponential growth"?

7) Cite 2 websites where the phrase "exponential growth" is used, read the particulars of the website, and state whether the example is actually showing exponential growth.

8) If you look at a given set of data, explain 2 ways to tell if the data is exponential.

9) Explain why the following functions are *not* exponential?

    (a) $H(s) = 3s - 9$

    (b) $Q(x) = 15x^2 - 2x + 1$

10) If bacteria, viruses, and molds grow exponentially, why is it that they don't take over the world? In other words, why do they eventually stop growing and die? ((You may need to do some research for this question. List all sources))

11) Explain why the base of an exponential function cannot equal 1.

12) Why does the base of an exponential function need to be greater than 0 (positive)?

13) What happens to the graph of any exponential function if the value of a is less than 0?

14) List as many properties of exponential functions as you can.

15) Given the general form of an exponential function, $f(x) = a \cdot b^x$, show that when the input value is 0, the output value is always the y-intercept. In other words, show that $f(0) = a$.

16) Describe the shape of exponential functions when the value of b is in the following ranges...

(a) $b < 0$

(b) $b = 0$

(c) $0 < b < 1$

(d) $b = 1$

(e) $b > 1$

17) What is the domain of the family of exponential functions, $f(x) = a \cdot b^x$ (with $a > 0$, $b > 0$, $b \neq 1$)?

18) What is the range of the family of exponential functions, $f(x) = a \cdot b^x$ (with $a > 0$, $b > 0$, $b \neq 1$)?

19) What is the range of the family of exponential functions, $f(x) = a \cdot b^x$ (with $a < 0$, $b > 0$, $b \neq 1$)?

20) What is the domain of the family of exponential functions, $f(x) = a \cdot b^x$ (with $a < 0$, $b > 0$, $b \neq 1$)?

21) What is an asymptote? How does it relate to the graph of an exponential function?

22) Given the statement, "The population of Scandinavian countries declines as a function of time," explain whether this growth is more likely to be linear or exponential.

23) As x increases, the graph of an exponential function gets closer and closer to the line y = 3, but it never actually touches the line. This is an example of what concept?

24) Translate the following math sentence into English... as $x \rightarrow \infty$, $f(x) \rightarrow 0$.

25) Explain the difference between the decay factor and the decay rate.

26) In your own words, give the definition of half-life.

27) Carbon 14 is used in dating many archeological finds (called Carbon Dating...by the way, you should never date Carbon). What is the half-life of C-14?

28) What is meant by *compound* interest?

29) Why does compound interest earn a larger profit than simple interest?

30) In general, which will earn you more money in your savings account, increasing the number of compounding or increasing the interest rate?

31) You have $5000 to invest in a CD which must be locked for 2 years, meaning you can neither deposit nor withdraw money. For each of the 4 accounts below, calculate the amount of money you will have after 2 years *and* the interest earned over the 2-year period.

(a) 3.5% compounded quarterly

(b) 3.5% simple interest

(c) 3.6% compounded quarterly

(d) 3.6% compounded continuously

32) The TI table below shows the amount of money (Y) growing in a *simple interest* investment over X years. Find the exponential regression equation and use that to determine (a) the Principal and (b) the interest rate.

X	Y1	
1	7837.5	
2	8190.2	
3	8558.7	
5	9346.4	
7	10206	
9	11146	

X=

33) *Loaded vs. no-load mutual funds*

   A mutual fund really represents of group of different stocks, bonds, corporations, etc. Since mutual funds are typically diversified (they don't put all of their eggs in one basket), they can offer better rates of return then savings accounts or CD's. Anywhere from 8% – 12% is a likely rate of return. A *loaded* mutual fund is one that you pay a fee to be able to invest. These fees typically range from 4% – 8% of your initial investment. Loaded funds are totally not cool because you end up losing money and there are always no-load alternatives available. But hey...you didn't know this did you? Calculate the earnings for the 2 investments below and describe what happens when you purchase a loaded vs. a no-load mutual fund.

No-load fund: You invest $10,000 at 8% compounded continuously for 7 years.

Loaded fund: You invest $10,000 at 8% compounded continuously for 7 years, but you need to pay 5% of your investment up front

34) On June 11, 2007, the following description of iLike™ was posted on the website listed below. The **bold** words are my own emphasis. (Source: http://mashable.com/2007/06/11/ilike-facebook-app-success/)

*iLike has gotten a reported three million users since its launch last October, and its growth is **exponential**, now that it's on Facebook. The music website is gaining nearly **a million new users every month**, and is looking to become the most used music application on Facebook.*

Does this situation of growth really describe an exponential function? Why or why not? If not, what family of functions does it describe (and why)?

35) Write the equations for two different exponential functions that increase from left to right and cross the y-axis at (0, 3). Call them f(x) and g(x).

36) Write the equations for two different exponential functions that decrease from left to right and cross the y-axis at (0, -2). Call them h(x) and p(x).

37) Write a possible equation for an exponential function that has a domain of all real numbers and a range of y < 0. Call it f(x).

38) Write a possible equation for an exponential function that has a domain of all real numbers, a range of y < 0, and a y-intercept at (0, -5). Call it T(m).

39) Write a possible equation for an exponential function that has a domain of all real numbers, a range of y > 0, and a y-intercept at (0, 12). Call it R(n).

40) An exponential function has a growth rate of 25%. What is the base of the function?

41) An exponential function has a growth rate of 58%. What is the base of the function?

42) An exponential function has a decay rate of 36%. What is the base of the function?

43) The base of an exponential function is 1.82. Is the function growing or decaying? Why? What is the rate of growth (or decay)?

44) The base of an exponential function is 0.64. Is the function growing or decaying? Why? What is the rate of growth (or decay)?

45) If the half-life of a radioactive compound is 125 years, what will happen every 125 years?

46) Write the equation of an exponential function that starts out at 1000 and grows 18% each year. Call it G(t).

47) Write the equation of an exponential function that starts out at 1000 and decays 18% each year. Call it D(t).

48) If the frequency of a note on a piano is measured at 945 Hz, what would you do to find the frequency of the next note (one half-step up)? Calculate the frequency.

49) Determine each of the following without using your calculator:

(a) $25^{3/2}$ (b) $81^{-1/2}$ (c) $64^{2/3}$

50) For the data below, find the exponential equation of best fit *algebraically* (by hand). Verify your answer using ExpReg on the TI.

Time (hours)	0	1	2	5
Number of infected people	150	188	234	458

51) In the following tables, determine whether the data is exponential, linear, or neither. If the data is exponential find the equation in $y = a \cdot b^x$ form. If the data is linear find the equation in $y = mx + b$ form. If the data is neither, write "neither."

(a)

X	Y1	
0	125	
2	92.45	
3	79.507	
4	68.376	
7	43.491	
X=		

(b)

X	Y1	
0	85	
1	87.45	
3	92.35	
5	97.25	
6	99.7	
7	102.15	
X=		

(c)

X	Y1	
0	42	
2	44	
3	51	
4	62	
8	146	
9	177	
X=		

(d)

X	Y1	
0	-20	
1	-46	
2	-105.8	
3	-243.3	
4	-559.7	
6	-2961	
X=		

52) Write the domain and range for the following exponential graphs:

(a)

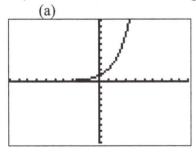

(b)

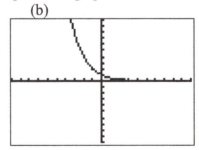

(c)

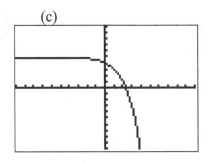

(d)

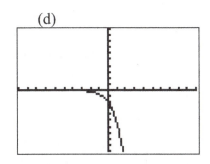

(e)

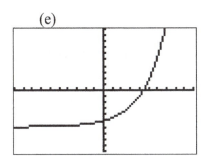

Application: Frequency of Vibration

53) Changing the length of a guitar string changes the frequency at which it vibrates and therefore changes the note. Playing 1/2 of the original string length plays a note an octave higher than the full string. The rate of vibration of this note is twice as fast as the original. If the original note was a 400 Hz tone, then the string of half length is producing an 800 Hz tone. Holding the string so that it is 2/3 its original length plays a note a *fifth* higher than the original note. The string is now vibrating 3/2 as fast as it was originally, so that if the original tone was 400 Hz, the fifth is 600 Hz.

The table below shows the note's relative position and ratio of vibration with respect to the original note.

Number of note played	Relative position to original note	Rate of vibration to original	Frequency of vibration (Hz)
1	Octave	2	800
2	Major 6th	5/3	666.66
3	Minor 6th	8/5	640
4	Fifth	3/2	600
5	Fourth	4/3	533.33
6	Major 3rd	5/4	500
7	Minor 3rd	6/5	480

    (a) Using the TI-83, find the exponential equation of best fit for Rate of vibration (column 3) as a function of note played (column 1). List your r-value.

    (b) Using the TI-83, find the exponential equation of best fit for Frequency of vibration (column 4) as a function of note played (column 1). List your r-value.

    (c) What do you notice about the two equations and the two r-values? How can you explain the relationship between these two equations?

    (d) Using the equation found in part (a), predict the rate of vibration for the 8th note played.

54) In the world of audio, the unit for power is the *watt*, and the unit for SPL or Sound Pressure Level (loudness) is the *decibel*. (A decibel is one-tenth of a bel, since deci is the metric prefix for one-tenth. The bel was actually named after the founder of the telephone, Alexander Graham Bell).

  An audio speaker is rated at 92dB per watt/meter. This means that 1 watt will produce 92dB at 1 meter away from the speaker (this is the method of efficiency rating a speaker). The table below shows the power output of the speaker for different decibel levels. (Notice that we carried loudness to two *decibel* places ☺)

SPL (decibels)	Power (watts)
92.00	1
95.00	2
96.75	3
98.00	4
98.97	5
99.75	6
100.42	7
101.00	8
101.51	9
101.97	10

a) Use the data to find the exponential regression equation for SPL as a function of power. Since this equation may be cumbersome with the scientific notation, have the TI place the equation into Y_1 automatically by using the method I showed you in the *Booklet on Linear Functions*.

(b) Use the Table on the TI to predict the power output needed for 110 dB.

(c) For humans the threshold of pain is 120 dB. Sound pressure level becomes lethal at about 130dB (note that the speaker could not produce 130 dB – it would be damaged way before that). How many watts would be required to produce 130 dB of volume?

(d) Round your answer in part (c) to the nearest thousand and do a google™ search to see what can produce this type of power. For example, if your answer was 34 watts, type in "30 watts" in google.

55) Ra-223 is an isotope of radium that is currently under investigation to use in medicine for some cancer treatments. The half-life of Ra-223 is approximately 11.4 days. If you have 10g of Ra-223 in your pocket – which, by the way, I do not advise – answer the following:

(a) In how many days (total) will there be 5g remaining?

(b) In how many days (total) will there be 2.5g remaining?

(c) Using the 10g as day 0, find the exponential equation that shows number of grams remaining as a function of the number of days.

(d) Use your equation to predict the number of grams of the isotope remaining after 100 days. What about after 1000 days?

(e) Does the radioactive isotope ever completely disappear? Why or why not?

56) According to CIA statistics as of mid 2007, India's population was approximately 1.129 billion, whereas China (the most populated country in the world) had a population of about 1.321 billion. India's growth rate is estimated to be 1.38% per year, whereas China's is 0.606% per year.

(a) Using 2007 as year 0, write an exponential equation for the population of India as a function of year. Call it I(x).

(b) Using 2007 as year 0, write an exponential equation for the population of China as a function of year. Call it C(x).

(c) Using the Intersect method or the Table method on your TI, predict in what year India will overtake China as the most populated country on the planet.

57) The following table shows the average price for 1 gallon of regular unleaded gasoline in the US from 1986 to 2007.

Remember to use the data trick that we learned!

Make 1986 year 0, 1987 year 1, etc…

YEAR	PRICE (cents)
1986	93
1987	95
1988	95
1989	102
1990	116
1991	114
1992	113
1993	111
1994	111
1995	115
1996	123
1997	123
1998	106
1999	117
2000	151
2001	146
2002	136
2003	159
2004	188
2005	230
2006	259
2007	275*

Source: www.fueleconomy.gov
2007 is estimated.

(a) Using your TI-83 find the exponential regression equation for this data.

(b) Use the equation from part (a) to predict the average price for 1 gallon of unleaded fuel in the US in 2010.

(c) How valid is your prediction? Consider the r-value in your response.

Booklet on Logarithmic Functions

The Scoville scale

The Scoville scale is a measure of the hotness of a chili pepper. These fruits of the Capsicum genus contain capsaicin, a chemical compound which stimulates chemoreceptor nerve endings in the skin. The number of Scoville heat units (SHU) indicates the amount of capsaicin present. Some hot sauces use their Scoville rating in advertising as a selling point. The scale is named after its creator, American chemist Wilbur Scoville, who developed a test for rating the pungency of chili peppers. His method, which he devised in 1912, is known as the Scoville Organoleptic Test. *(Source: www.wikipedia.org)*

The table below shows eleven species of peppers, the average Scoville rating for members of that species, and a numerical listing from least to most hot.

Species of pepper	Scoville rating (x)	Numerical listing of hotness (y)
Bell	1	1
Pepperoncini	300	2
Anaheim	1500	3
Jalapeño	5,000	4
Serrano	15,000	5
Cayenne	40,000	6
Thai	75,000	7
Jamaican hot	150,000	8
Habanero chili	225,000	9
Red Savina Habanero	450,000	10
Naga Jolokia	900,000	11

CAPSAICIN

The molecule above is capsaicin, the main ingredient in chili peppers. Different forms of capsaicin produce different levels of "irritation," and therefore have a higher Scoville rating.

Using the Scoville rating as the input variable and the numerical listing as the output variable, create a scatter plot using your TI-83. Be sure to adjust the window so that all points fit in your plot. Sketch the scatter plot below…

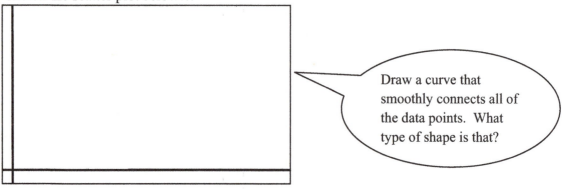

Draw a curve that smoothly connects all of the data points. What type of shape is that?

Your scatter plot should look something like the plot below, if you choose a window as I have chosen (or at least close to mine).

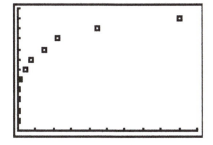
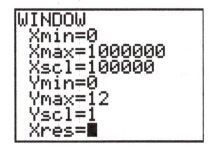

What shape is this?

If you drew your smooth curve over the data points on the previous page, congratulations!! You may have drawn your first *logarithmic* function ever!

Characteristics of the logarithmic family of functions

Logarithmic functions initially rise very quickly, but then the increase begins to slow as x moves further to the right. Make no mistake, they do not "flatten out." Logarithmic functions continue increasing forever, however the rate at which they increase is always decreasing. The shape of a logarithm is *similar* to the square root function, but it has some distinct differences...

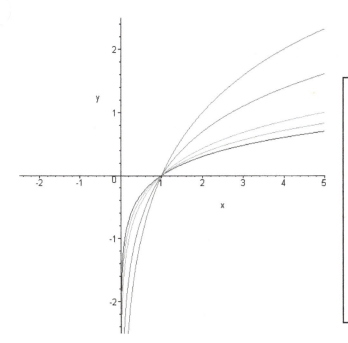

Logarithmic functions do the following...

1) They approach the y-axis as x gets closer to 0. In fact, the y-axis is an asymptote for the functions.

2) As x increases, the function increases. Although the function increases more slowly as x moves further to the right.

3) They have no y-intercept, but cross the x-axis once...typically at (1, 0).

In your mind, pick one of those curves and flip it over the line y = x, sketch what it might look like on the graph above. What type of function did you just draw?

MENTAL EXERCISE

if you guessed *exponential*, you are correct again. It is a well known fact that…

Logarithmic and exponential functions are inverses.

The TI screen below shows graphs of $f(x) = 10^x$, $h(x) = x$, and $g(x) = \log_{10}x$

$f(x) = 10^x$

$h(x) = x$

$g(x) = \log_{10}x$

MEMORY BREAK

If you need a small refresher on the characteristics of inverse functions, go back to the *Booklet on Functions*

Looking at the graph above, we can see the following inverse relationships…

(1) $f(x)$ and $g(x)$ are reflections over the *mirror* line $y = x$.

(2) The point $(0, 1)$ is on $f(x)$ and the point $(1, 0)$ is on $g(x)$.

(3) The x-axis is an asymptote of $f(x)$, while the y-axis is an asymptote for $g(x)$.

(4) $D_f = R_g$, namely all real numbers.

(5) $R_f = D_g$, namely all positive real numbers.

This means that the inverse of b_x is $\log_b x$…and vice-versa.

Examples:

(a) If $f(x) = 2^x$, then $f_{inv}(x) = \log_2 x$.

(b) If $f(x) = e^x$, then $f_{inv}(x) = \log_e x$

(c) If $f(x) = 15^x$, then $f_{inv}(x) = \log_{15} x$

REMINDER…
   Most texts use the notation, $f^{-1}(x)$ to represent *the inverse of f(x)*. I believe a better (less confusing) notation was to use $f_{inv}(x)$ to represent *the inverse of f(x)*.

MATH 2 ENGLISH

How do we say…

$\log_b x$	"log base b of x"
$\log_2 x = y$	"log base 2 of x is y"

Knowing that, we can make the following generalizations given an exponential function, b^x, and its inverse (logarithmic) function, $\log_b x$...

(a) $b^{\log_b(x)} = x$. In other words $f(g(x)) = x$

(b) $\log_b(b^x) = x$. In other words $g(f(x)) = x$

This was from our composition definition of an inverse...$f(g(x)) = g(f(x)) = x$ *iff* $f(x)$ and $g(x)$ are inverse functions.

Professor's Practice Pause

(1) List as many characteristics of logarithmic functions as you can.

(2) State several algebraic or graphical results which show that logarithmic functions and exponential functions are inverses.

(3) Write the inverse of the given functions by inspection.

    (a) 6^y (b) $\log_3 x$ (c) 10^z

    (d) b^x (e) $\log_b x$ (f) $\log_e x$

(4) Using the composition definition of an inverse, show that 5^x and $\log_5 x$ are inverses.

Associated End of Booklet exercises are 1 – 5

$\mathcal{BEN}$ notation

While it is true that exponential and logarithmic functions are inverses, it is also true that logarithms <u>are</u> exponents. What do I mean by that??

The answer to a logarithmic expression is an exponent.

In order to show this wonderful little fact, we introduce our friend, BEN. BEN is your friend! If you don't believe me, take a look at that smiley face up above.*!!* BEN stands for **B**ase, **E**xponent, and **N**umber. It's the form we use to write all of our simple exponential equations.

For example, we usually write exponential equations like this…

$$3^2 = 9$$

$$\text{Base}^{\text{Exponent}} = \text{Number}$$

The inverse of that relationship gives us a very important tool in helping to understand logarithms and in solving logarithmic equations.

$$b^E = N \leftrightarrow \log_b N = E$$

Notice that the answer to the logarithm equation (on the right) is the exponent of the first expression (on the left)…hence, logarithms are exponents.

So let's put some numbers in there to see if we can pick up the pattern.

$3^2 = 9$……..is equivalent to saying……….$\log_3 9 = 2$

Note: A way to say this in words would be "What number do I have to put on top of a 3 in order to come up with 9?....answer: 2, because 3 squared is 9."

$4^2 = 16$……….. is equivalent to saying……….$\log_4 16 = 2$
$5^3 = 125$…….. is equivalent to saying………..$\log_5 125 = 3$
$10^2 = 100$……. is equivalent to saying……….$\log_{10} 100 = 2$

The logarithm base 10 is called the "<u>common logarithm</u>" since we do everything in base 10 (counting, money, metric system, etc.) Because mathematicians are basically lazy, we drop the 10 and just write **log100 = 2**. So if there is no base, we understand that to mean base 10.

SHORTCUT 1: $\log x$ is a shortcut for $\log_{10} x$

Mathematical grammar note...We can either use parentheses when writing the argument of a logarithm or not. For example, either of these is okay: $\log_5 x$ or $\log_5(x)$

~ An Interesting Thought ~

Why do people count in base 10? Ancient Babylonians counted using base 60 for reasons of commerce (or maybe time), but essentially all people groups have counted using base 10. Any ideas why? If you're not sure, start counting as if you were 4 years old again...

Continuing...

$6^1 = 6$..........corresponds to saying..........$\log_6 6 = 1$

$10^1 = 10$.......corresponds to saying..........$\log 10 = 1$

$$\log_b(b) = 1$$
...because
$$b^1 = b$$

The last two examples show that the *logarithm base "b" of "b" always equals 1*, because any number raised to the 1 is itself.

$2^0 = 1$..........corresponds to saying..........$\log_2 1 = 0$

$10^0 = 1$.........corresponds to saying.........$\log 1 = 0$

$$\log_b(1) = 0$$
...because
$$b^0 = b$$

The previous two examples show that the *logarithm any base of 1 is 0*, because anything raised to the 0 power is 1.

$e^0 = 1$..........corresponds to saying..........$\log_e 1 = 0$

$e^1 \sim 2.718$..........corresponds to saying..........$\log_e 2.718 \sim 1$

The logarithm base e is called the "natural logarithm", since the number e is referred to as the natural base. Remember we said that e shows up in pretty much all of the sciences, economics, etc. We also discussed that the symbolism most likely came from the Swiss Mathematician, Leonhard Euler, who spoke French. Monsieur Euler would not have said "natural logarithm" because that's English. He would have most likely said "le Logarithme Naturel" which is (perhaps) where the "ln" comes from. There are other ideas about that one, though...

SHORTCUT 2: $\ln x$ is a shortcut for $\log_e x$

Notice that these are the two buttons that appear on your calculator, next to the "7" and the "4". While there are some applications using logarithms of different bases, the vast majority of applications use either the common or the natural logarithm. Let's break it down...

Name :	Common log	Natural log
Math:	$\log_{10} x$	$\log_e x$
Shortcut: (always used)	$\log x$	$\ln x$

Professor's Practice Pause

(1) Change all exponential equations below to their corresponding logarithmic equations using the BEN relationship. Use shortcut notation for base e and base 10.

(a) $5^2 = 25$

(b) $3^0 = 1$

(c) $(1/2)^3 = 1/8$

(d) $e^1 = e$

(e) $10^4 = 10,000$

(f) $10^{-2} = .01$

(2) Change all logarithmic equations below to their corresponding exponential equations using the BEN relationship. Keep in mind the shortcut notation for base e and base 10.

(a) $\log_4 64 = 3$

(b) $\log_2 64 = 6$

(c) $\log_{1/3} 9 = -2$

(d) $\ln 1 = 0$

(e) $\log x = y$

(f) $\log 10 = 1$

(3) Evaluate the following expressions *without* the use of your calculator:

(a) $\log 1,000,000$

(b) $\log_{1/3}(3)$

(c) $\ln 1$

(d) $(1/2)^{-2}$

(e) $\ln e$

(f) $\log(.0001)$

(4) Use the BEN relationship to solve the following equations:

(a) $\log_2(x) = 4$

(b) $\log(y) = 2$

(c) $\log_b x = 0$

(5) Sketch the graphs of $f(x) = \log x$ and $g(x) = \ln x$ from your calculator in the space provided on the next page.

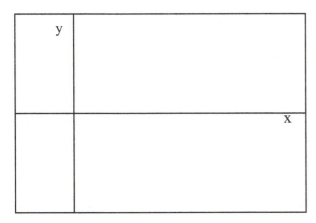

Which of the two logarithmic functions rises faster? Slower? Use the inverse functions to help explain why this is the case.

Associated End of Booklet exercises are 6 – 8.

Properties of logarithms

Below is a summary of the logarithmic properties used to "expand" and "contract" logarithmic expressions. Sometimes we need to expand the logarithmic expressions (going from left to right in the equations below), and sometimes we need to contract them (going from right to left).

Product property

$$\log_b(MN) = \log_b(M) + \log_b(N)$$

"The logarithm of a product is the sum of the logarithms"

Quotient property

$$\log_b\left(\frac{M}{N}\right) = \log_b(M) - \log_b(N)$$

"The logarithm of a quotient is the difference of the logarithms"

Exponent property

$$\log_b(x^n) = n \log_b(x)$$

"The logarithm of something raised to a power is the power times the logarithm"

Change of base property

$$\log_b(x) = \frac{\log_a(x)}{\log_a(b)}$$

"The logarithm of a quantity in an old base is equal to the logarithm of the quantity in the new base divided by the logarithm of the old base."

EXAMPLE

Expand the following expression to an expression with multiple log terms. $\log\left(\dfrac{10\,x^3\,y^2}{2\,z^5}\right)$

Using property 2, we can change this to be $\log(10\,x^3\,y^2) \,-\, \log(2\,z^5)$

Then, using property 1, the expression becomes…

$\log(10) + \log(x^3) + \log(y^2) - [\log(2) + \log(z^5)]$ which is

$\log(10) + \log(x^3) + \log(y^2) - \log(2) - \log(z^5)$ after distributing the negative

Notice every factor that came from the *numerator* has a + sign on the associated log term, while every factor that came from the *denominator* has a – sign in front….that's a short cut.

But we are not yet done….

Using property 3, we can simplify further…

$\log(10) + 3\,\log(x) + 2\,\log(y) - \log(2) - 5\,\log(z)$

We know that $\log(10) = 1$….remember no base means base 10.

Also, the logarithm of a number is just another number. $\log(2)$ on your calculator gives 0.301, so we combine the numbers to get 0.699

$.699 + 3\,\log(x) + 2\,\log(y) - 5\,\log(z)$ This is the *expanded* form.

Note that we changed the logarithm of a complicated polynomial and simplified it to "linear" type functions of logarithms.

Use this space for Example 2, done in class. *Contract* $2 \ln(x) - 3 \ln(y) + \ln(x^2) - .5 \ln(z)$

Use this space for Example 3, done in class. *Expand* $\log_2\left(\dfrac{x^4 y^3}{z}\right)^{\left(\frac{1}{3}\right)}$

Use this space for Example 4, done in class.

Change $\log 7(x)$ to base 10

Change $\log 7(x)$ to base e

Change $\log_5(x^2)$ to base 2

Professor's Practice Pause

Using the 4 properties of logarithms

Expand each of the following logarithmic expressions to an expression with multiple terms.

(a) $\log\left(\dfrac{P^3 Q}{M^2}\right)$

(b) $3 \ln(4 x z^3)^2$

(c) $\log_4\left(\dfrac{64\, n^2}{m^3\, s}\right)$

Contract each of the following logarithmic expressions to an expression with a single term.

(a) $\ln(x) + 3 \ln(x^2) - 4 \ln(y)$

(b) $2 \log(A) - 5 \log(A)^2 - 7 \log(A)^3$

Change all of the following expressions to both base 10 and base e.

(a) $\log_3(C)$

(b) $\log_\pi(x)$

(c) $2 \log_2(x^2)$

Associated End of Booklet exercises are 9 – 14.

Solving exponential and logarithmic equations

Example 1: Algebraically solve $180 = 11.80(1.077)^t$

1. We first ISOLATE the exponential factor...

$$\frac{180}{11.8} = \frac{11.80}{11.8}(1.077)^t$$

$$15.254 = 1.077^t$$

2. Now we "take the ln" of both sides...as follows...
     (we could also use log base 10)

$$\ln(15.254) = \ln(1.077)^t$$

3. Use property 3 to bring the t down in front...

$$\ln(15.254) = t * \ln(1.077)$$

4. Isolate the t by dividing both sides by $\ln(1.077)$.   Remember, *"the logarithm of a number is just another number."*

$$t = \ln(15.254) / \ln(1.077) \sim 36.7$$

Professor's Practice Pause

Given $\log(x + 3) + \log(x + 4) = 1$

Solve this equation using the Intersect method on the TI-83. That is, put the left side of the equation in Y1, the right side in Y2...and press 2nd, Trace, Enter, Enter, Enter.

Write your solution here: _____

See the next page for the solution...

Associated End of Booklet exercise is 15.

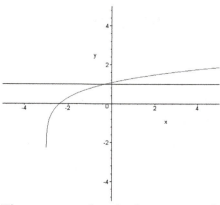

The answer using the intersect method is <u>x = -0.298</u>

BUT…how do we solve this equation ALGEBRAICALLY..?

Example 2:

1. Start with the original…log(x+3) + log(x + 4) = 1

2. Use the properties to *contract* to one log…

$$\log_b(MN) = \log_b(M) + \log_b(N)$$

log[(x + 3)(x + 4)] = 1

3. Use BEN notation to change the equation around…

$10^1 = 10 = (x + 3)(x + 4)$

$$x = -\frac{b}{2a} \pm \frac{\sqrt{b^2 - 4ac}}{2a}$$

4. Foil and subtract the 10 to get….

$0 = x^2 + 7x + 2$

5. Although we will encounter the *quadratic formula* in the next Booklet, you should remember this from your prerequisite math course!! Using the QF, you get x = **-0.298** and x = -6.701. We would throw away the -6.701 answer because when you put it back into the original equation you would get the log of a negative number…which does not exist. Carefully consider the check below…

CHECK…x = -0.298

log(-0.298+3) + log(-0.298 + 4) = 1
log(2.702) + log(3.702) = 1
log(2.702·3.702) = 1
log(10) = 1…..√

CHECK…x = -6.701

log(-6.701 + 3) + log(-6.701 + 4) = 1
log(-3.701) + log(-2.701) = 1

NO!!

The domain of log(x) is x > 0

Example 3:

Use **BEN** to solve logarithmic equations…

Solve… $3 \log_5(x + 2) = 5$

1. Isolate the logarithmic term by dividing both sides by 3…

$$\frac{3 \log_5(x + 2)}{3} = \frac{5}{3}$$

$$\log_5(x + 2) = 5/3$$

2. Use BEN to rewrite the equation as an exponential equation

b is 5, N is $(x + 2)$, and E is $5/3$

$$5^{(5/3)} = x + 2$$

3. Subtract 2 from both sides to isolate x…

$$x = 5^{(5/3)} - 2 = \mathbf{12.62}$$

CHECK:

$3 \log_5(12.62 + 2) = 5 \longrightarrow 5 = 5……√$
$3 \log_5(14.62) = 5$
$3 \dfrac{\log 14.62}{\log 5} = 5$

Example 4:

Use change of exponential bases to solve an exponential equation.

Solve $2^x = 16^{(3 - x)}$

The goal is to re-write the 16 as some form of 2, or the 2 as some form of 16. It will be easier to follow if we turn 16 into 2^4. We can then rewrite the original equation as

$$2^x = \left(2^4\right)^{(3-x)}$$

$$2^x = 2^{(12 - 4x)} \quad …..\text{using our property of exponents } (x^\wedge m)^\wedge n = x^\wedge(mn)$$

Now the key…**If the bases are equal, then the exponents must also be equal.**

$x = 12 - 4x$, which leads to $\mathbf{x = 12/5}$…..I leave the check up to you.

Professor's Practice Pause
Solving Exponential and Logarithmic equations algebraically

Solve the following equations for x. Be sure to verify your solutions algebraically and using the Intersect method on the TI.

1) $3\,e^x - 7.64 = 1.3$

2) $6\log_3(2\,x - 10) = 7$

3) $\ln(5) - \ln(x + 1) + \ln(x) = 0$

4) $\left(\dfrac{1}{3}\right)^x = 27^{(1-x)}$

Associated End of Booklet exercises are 16 and 17.

Modeling Logarithmic Functions (LnReg)

…Back to the Scoville scale…

In Class Exercise: Enter the Scoville number vs. Numerical listing of hotness data into your calculator as you did at the beginning of this Booklet. For simplicity, enter the input data (Scoville rating) in L_1 and the output data (listing of hotness) into L_2.

> **Author's note**…and might I say, Professor's note as well…
>
> At this point in the course, you should be able to do the steps involved with regression analysis in your sleep. If not, we (that is me and your professor) recommend that you go back to the *Booklet on Linear Functions* and re-read the section entitled *Modeling linear functions on the TI-83*, on or around page 190 (depending on the edition of this text). Specifically at this point in time, you should be very comfortable with the following:
>
> (a) Entering data into lists
> (b) Generating a scatter-plot
> (c) Choosing a regression model
> (d) Having the calculator find the regression equation
> (e) Using that equation to make predictions (interpolation or extrapolation)

Find the logarithmic regression equation. That is the logarithmic equation that best fits the data. In the STAT / CALC menu, you will want to choose number **9: LnReg**.

Notice that the form of the equation is $y = a + b \cdot \ln(x)$. The calculator always uses the natural logarithm for regression.

(a) Write your answer here _____

(b) Write your r-value here _____

(c) Is the r-value good enough to make predictions? Explain.

(d) A new hybrid pepper is created that has a Scoville rating of 650,000. Use your equation to predict the hotness listing of this pepper. Does the answer fit the data? Why or why not?

(e) You friend says she has created a batch of chili that has a numerical listing of hotness equal to 12.4. Use the methods of this booklet and your equation to solve for the corresponding Scoville rating. In other words, solve the equation $12.4 = -1.229 + 0.765 \cdot \ln(x)$.

(f) Solve the equation if the hotness listing was 8.75, instead of 12.4

(g) Solve the equation if the hotness listing was 5.39.

(h) You should see a pattern emerging in the way you solved the equations from questions (e) – (g). In the space below, solve the equation $A = B + C \cdot \ln(x)$

Group Pause

The prime counting function

As you recall from the *Booklet on Real numbers*, a prime number is a positive integer that has two factors, itself and 1. The first prime number is 2 (which is also the only even prime, since all other even numbers are divisible by 2). A number that is not prime is *composite*. The number 1 is neither prime nor composite, it is the unit number.

(1) Circle all the prime numbers below. You can look back to the Practice Pause in the first Booklet to check your answers.

1 2 3 4 5 6 7 8 9 10 11 12 13 14 15 16 17 18

19 20 21 22 23 24 25 26 27 28 29 30 31 32 33 34 35

36 37 38 39 40 41 42 43 44 45 46 47 48 49 50

Mathematicians define the *prime counting function*, $\pi(x)$, as the number of prime numbers less than or equal to x. For example, $\pi(1) = 0$ because there are no prime numbers less than or equal to 1. $\pi(2) = 1$, $\pi(3) = 2$, $\pi(4) = 2$, and so on.

(2) Using the circled numbers above, complete the following table.

x	$\pi(x)$
1	0
2	1
3	2
5	
10	
20	
50	

(3) Create a scatter plot of the above data, and find the logarithmic function of best fit using the LnReg function. Write the equation and r-value.

(4) Is the equation a good match for the data? Use the equation to estimate $\pi(100)$. What is the actual answer?

Associated End of Booklet exercises are 18 – 20.

Topical Summary of Logarithmic Functions

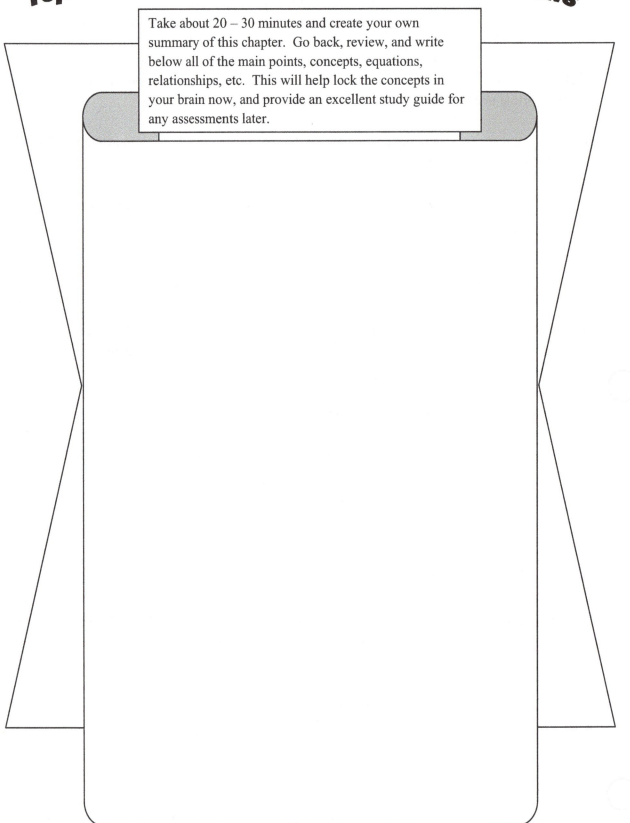

Take about 20 – 30 minutes and create your own summary of this chapter. Go back, review, and write below all of the main points, concepts, equations, relationships, etc. This will help lock the concepts in your brain now, and provide an excellent study guide for any assessments later.

End of Booklet Exercises

(1) Use your TI-83 to graph the functions f(x) = log(x) and g(x) = ln(x). Sketch and label these graphs below. Describe the interesting features of the graphs, including intercepts, asymptotes, etc.

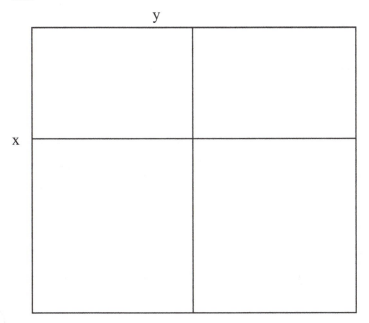

(2) State several reasons why exponential and logarithmic functions are inverses.

(3) Given the function f(x) = log$_b$(x), where b > 0, state the domain and range of this function.

(4) For the given functions, write the equation of the inverse by inspection.

   (a) $\log_6(x)$ (b) $\ln(x)$ (c) 10^x (d) π^x

(5) Use the composition definition of an inverse to show that f(x) = 5·2^x and g(x) = log$_2$(x/5) are inverse functions

(6) Using *BEN* notation, re-write the following equations into their exponential or logarithmic equivalents. Use the shorthand notation for base 10 and base e.

(a) $2^4 = 16$

(b) $15^2 = 225$

(c) $10^4 = 10000$

(d) $8^{1/3} = 2$

(e) $\sqrt{64} = 8$

(f) $e^0 = 1$

(g) $\ln(x) = 5$

(h) $\log(100) = 2$

(i) $\log_3 81 = 4$

(j) $\log(1) = 0$

(k) $\ln(1) = 0$

(l) $\log_5(1) = 0$

(7) Without the use of a calculator, evaluate the following expressions.
  (a) $\log 1000$

  (b) $\log(.001)$

  (c) $\ln(e)$

  (d) $\log_5(0)$

  (e) $\log_{1/5}(25)$

  (f) $(1/3)^{\wedge}(-3)$

(8) Use the BEN relationship to solve the following equations.
  (a) $\log_3(x) = 27$

  (b) $10^x = 2.5$

  (c) $\ln(2x - 4) = 1$

9) Use the properties of logarithms to *expand* the following expressions.

(a) $4 \log_7 \left(\dfrac{x\,y}{z^2} \right)$

(b) $\log \left(\dfrac{100\, p\,(p+1)}{p^2\,(2-p)} \right)$

(c) $(\ln(z^2))^2$

(10) Use the properties of logarithms to *contract* the following expressions.

(a) $2 \log(x) - 5 \log(y) + \log(z^2)$

(b) $\ln(5\,x) + \ln(x^2) - \ln(e) - \ln(6\,x^3)$

(c) $\log_5(x) - 2 \log_5(x^3) + 4 \log_5(3 - x)$

(d) Change your answer to part (c) into an expression in base 10

(11) Change your answers in Question 9 (parts (a) – (c)) to expressions in base 2

    (a)

    (b)

    (c)

(12) Prove the *Exponent Property* using the *Multiplication Property*.
    Hint: rewrite x^n as $x \cdot x \cdot x \cdot x \cdot \ldots \cdot x$ (n times).

(13) Given $\log\left(\dfrac{A\,B\,C\,D}{R\,S\,T\,V}\right)$. Use the *Product* and *Quotient* Properties of logarithms to expand the expression to multiple logarithmic terms. Also show that any term originally in the numerator is positive when expanded, and any term originally in the denominator is negative when expanded.

(14) Use the *Change of base* Property to show that $\log(x) = \ln(x)^{\left(\ln(10)^{(-1)}\right)}$

(15) Use the Intersect Method on the TI to solve each of the following equations. Check your answers algebraically.

(a) $3 \log(x + 1) = 5 \ln(x) - 1$ TI solution(s) _____

(b) $\log(x - 2) + \log(x - 1) = .5$ TI solution(s) _____

(c) $2^x = 4^{(3.15 - x)}$ TI solution(s) _____

(16) Solve the following equations algebraically and check your answers.

(a) $.68 \ln(x) - 2.35 = .1$

(b) $\log(3) - \log(x + 1) + \log(x) = 0$

(c) $2 \log_8(3x - 5) = 4$

(d) $\left(\dfrac{1}{4}\right)^{\left(\frac{x}{2}\right)} = 64^{(-x-2)}$

(e) $A \log(Bx + C) = D$

(17) How do we change exponential bases? For example, how do we go from an exponential expression in base 2 to an exponential expression in base e? Given $2^x = e^?$, solve for ?.

(18) The following table shows the average price for 1 gallon of regular unleaded gasoline in the US for various years from 1986 to 2007. Use price as the *input* variable and *YEARS from 1986* as the output variable.

PRICE (cents)	YEAR	YEARS from 1986
93	1986	0
95	1987	1
102	1989	3
111	1993	7
111	1994	8
115	1995	9
123	1996	10
123	1997	11
151	2000	14
146	2001	15
159	2003	17
188	2004	18
230	2005	19
259	2006	20
275*	2007	21

Source: www.fueleconomy.gov
* estimated.

(a) Using your TI-83 create a scatter plot for this data.

(b) Find the logarithmic regression equation of best fit.

(c) Use the equation from part (b) to predict the year in which gas will hit $4 per gallon (i.e. 400 cents)

(d) How valid is your prediction? Consider the r-value in your response.

(19) In this Booklet, we looked at the *prime counting function*, $\pi(x)$. A better approximation for $\pi(x)$ is $x/\ln(x)$. Use this approximation to calculate $\pi(100)$ and compare it with the answer you found in part (4) of the *Group Pause*.

(20) Some recent manufacturers of hot sauce have put out "pure capsaicin" products (with stringent warnings on the bottle I might add). The manufacturers claim that several of these products have up to 16,000,000 Scoville units. Using your regression equation for the Scoville scale, calculate the hotness listing of such a product.

Booklet on the Quadratic family of functions

Below is a summary table of the polynomial families that we will be studying for the next few Booklets. You may not understand the meaning of all of the columns right now, but that's okay. As we progress to the end of the Booklet on Higher Order Polynomial functions, I think you will find this table quite useful as a summary.

"Roots" is another name for solutions, as we will

The Polynomial Family of Functions

Polynomial Name	General Equation	General Shape	Number and Type of Roots
		(graphs show typical shapes for each family)	
Quadratic (Parabola)	$ax^2 + bx + c = 0$ (a, b, and c are real numbers)		2 roots 2R or 2C
Cubic	$ax^3 + bx^2 + cx + d = 0$ (a, b, c, and d are real numbers)		3 roots 3R or 1R, 2C
Quartic	$ax^4 + bx^3 + cx^2 + dx + e = 0$ (a, b, c, d, and e are real numbers)		4 roots 4R or 2R, 2C or 4C
Quintic	$ax^5 + bx^4 + cx^3 + dx^2 + ex + f = 0$ (a, b, c, d, e, and f are real numbers)		5 roots 5R or 3R, 2C or 1R, 4C

The R stands for real roots and the C stands for complex roots.

The Parabola - graph, vertex, and characteristics

The Arrow and The Song (an excerpt)

by Henry Wadsworth Longfellow

"I shot an arrow into the air,
It fell to earth, I knew not where;
For, so swiftly it flew, the sight
Could not follow it in its flight."

Do not try this at home, especially if you live in the suburbs.

Old Mr. Longfellow probably did not know it, but if he measured the arrow's distance above the ground versus time, he would have charted a _parabola_. In fact, anything that you throw, toss, chuck, shoot, lob, eject, or otherwise hurl into the air will chart a parabola if you graph the distance above the ground versus time. A parabola is simply a U-shaped curve that happens to be one of some things we call _conic sections,_ or "slices of a cone."

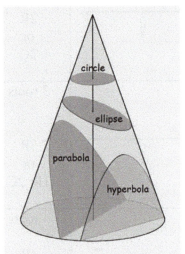

Apollonius of Perge, a Greek geometer who studied Euclid's _Elements_, is credited with the seminal work on conic sections. His greatest work, the _Conics_, is a beautiful and intricate compilation that has formed the basis of many geometry texts today.

In the figure to the left, slicing a cone parallel to the base will create a circle. If you angle that slice, but still cut through both sides of the cone, you will create an ellipse. Slicing a cone parallel to one side of the cone yields a parabola, while slicing perpendicular to the base of the cone creates a hyperbola.

Source: cseligman.com/text/history/ellipses.htm

Parabola and Quadratic are two sides of the same coin.

The parabola and the quadratic are like Jekyll & Hyde, Laverne and Shirley, or Abbott and Costello. ((You may have to look those up if they don't strike a chord of remembrance)). Basically, you can't have one without the other...Ahhh...remember the 90's sitcom "Married with Children"? The theme song was a version of _Love and Marriage_, performed by Frank Sinatra (please say you know him!). Anyway, the end of the chorus goes..."you can't have one without the other!"

The graph of a quadratic is always a parabola. The equation of a parabola is always a quadratic. The general equation for all members of the quadratic family of functions is...

$$f(x) = ax^2 + bx + c$$

We call ax^2 the **quadratic** term, bx the **linear** term, and c the **constant** term.

The constants a, b, and c are in $\mathbb{R}$, and $a \neq 0$. Why? What would happen if the constant a was allowed to be 0? We would have $f(x) = 0x^2 + bx + c = bx + c$, which is a line.

Examples of quadratic functions…

$f(x) = x^2$ $\qquad$ $g(x) = .75x^2 - x + 17$

$$y = -2x^2 - 4x$$

$\qquad y = -5x^2 + 1$ $\qquad\qquad$ $p(z) = 10z^2 + 5z + 4$

$f(t) = -(1/2)gt^2 + V_0t + H_0$

$\qquad\qquad\qquad\qquad y = (x - 3)(x + 2)$

$\qquad g(s) = (s - 2)^2 + 7$

What test do they pass?

When you graph a quadratic function, you get a parabola. Below are some examples of parabolas graphed using the TI-83. Notice that all parabolas are functions.

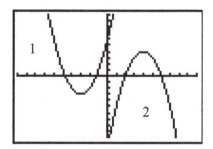
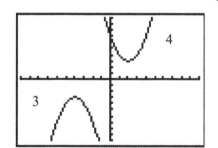

What we have learned so far is that this… \ / …is the graph of this…$ax^2 + bx + c$

Parabolas have several interesting characteristics (unlike lines which only have 2, slope and intercept). Although you may have done this already in high school with the almost, ever-present Mr. Gurglesteen, we will now study the *vertex* and *line of symmetry* of a parabola, and then take a look at some *characteristics*.

Vertex and line of symmetry

The vertex of the parabola is also called the "turning point" of the parabola. If the parabola "opens up" (parabolas 1 and 4 above), the vertex is the lowest point (the minimum). If the parabola "opens down" (parabolas 2 and 3 above), the vertex is the highest point (the maximum). The vertical line that passes through the vertex is called the line of symmetry; it divides the parabola into two, symmetrical halves.

Example: If the vertex of a parabola is the point (-2, 5), the line of symmetry is the line x = -2.

Example: If the vertex of a parabola is the point (4.65, -1.38), the line of symmetry is the line x = 4.65.

Example: Find the vertex and the line of symmetry of the parabola graphed below.

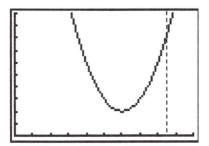

If we count tick marks, we can see that the vertex is the point (6, 2). The line of symmetry is then x = 6.

The line of symmetry is always
  x = the x-coordinate of the vertex

Can we find the vertex algebraically (using the equation)?

Yes, we can. Given $f(x) = ax^2 + bx + c$, the x-coordinate of the vertex is the point $-b/(2a)$. To find the y-coordinate of the vertex, we plug the x-value back into the function, so the vertex can be written as the point...

$$[-b/2a, \ f(-b/2a)]$$

Example: Given the quadratic function $g(t) = -3t^2 - 4t + 1$, find the vertex and the line of symmetry.

First we identify a and b. a = -3 and b = -4. The t-coordinate of the vertex is $-b/2a$ = -(-4)/(2·-3) = 4/-6 = -2/3.

To find the y-coordinate of the vertex, we calculate $g(-2/3) = -3(-2/3)^2 - 4(-2/3) + 1$ = -3(4/9) + 8/3 + 1 = -12/9 + 24/9 + 9/9 = 21/9 = 7/3.

The vertex is the point (t, g(t)) = (-2/3, 7/3). The line of symmetry is the vertical line t = -2/3.

Example: Given the quadratic function $f(x) = x^2 + 3$, find the vertex and the line of symmetry.

Since a = 1 and b = 0, -b/2a also equals 0. $f(0) = 0^2 + 3 = 3$, so the vertex is (0, 3). The line of symmetry is the y-axis (x = 0).

Professor's Practice Pause

For questions (1) through (5), state whether or not the given functions are quadratic. If not, explain why.

(1) $f(x) = 3 - 4x + 2x^2$

(2) $Z(d) = d^3 - 3d + 1$

(3) $y = 5x - 2$

(4) $y = -ax^2 - bx - c$

(5) $f(x) = 4x^2 - 2x^{1/3} + 5$

(6) Given $f(x) = -3x^2 + 5x - 1$. Graph this quadratic function on your TI, sketch the parabola below, and use your graph to estimate the vertex.

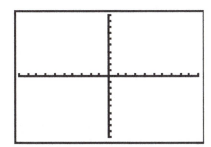

Estimate of vertex...

$$(\quad , \quad)$$

Now calculate the vertex exactly and compare it to your estimate.

Associated End of Booklet exercises are 1 – 5

From our understanding of how the vertex is found using the equation, it is clear that the letters a, b, and c must have something to do with the characteristics of the parabola. They do! Below is a group exercise that will help you determine what role they play…

Group Pause

In groups of 2 or 3, graph the following quadratic functions using your TI and try to make some determinations about the roles of a, b, and c in the general form of the quadratic $ax^2 + bx + c$.

The role of a: Graph the quadratics below either individually, in groups, or all together and use the graphical information along with the different values of a in each function to write several sentences about what the constant a does to the parabola.
$f(x) = -3x^2$, $g(x) = 3x^2$, $h(x) = -.5x^2$, and $p(x) = .5x^2$

The role of b: Graph the quadratics below either individually, in groups, or all together and use the graphical information along with the different values of b in each function to write several sentences about what the constant b does to the parabola.
$f(x) = x^2 + 4x + 1$, $g(x) = x^2 - 3x + 1$, $h(x) = x^2 - 6$, and $p(x) = x^2 - 9$

The role of c: Graph the quadratics below either individually, in groups, or all together and use the graphical information along with the different values of c in each function to write several sentences about what the constant c does to the parabola.
$f(x) = x^2 - 4x + 6$, $g(x) = -2x^2 - 8$, $h(x) = x^2$, and $p(x) = -.63x^2 + 3x - 3$

Associated End of Booklet exercises are 6 – 10

Now that we know some things about the vertex, line of symmetry, y-intercept, and other characteristics of a parabola, we should be able to describe quite a bit about a given parabola just by looking at its' corresponding quadratic function.

Example: Given the function $g(t) = -16t^2 + 16x + 80$, describe as much as possible about the parabola and then use your TI to graph the quadratic in an appropriate window.

We know that the parabola opens down because a < 0. Since |a| is > 1, the shape of the parabola is more narrow than wide, meaning it goes upward quite quickly. We also know that since b ≠ 0, the vertex is not on the y-axis. Finally, we know that the y-intercept is the point (0, 80) since $g(0) = -16(0)^2 + 15(0) + 80 = 80$.

We can calculate the vertex…$-b/2a = -16/(2 \cdot -16) = -16/-32 = 0.5$. $g(.5) = -16(.5)^2 + 16(.5) + 80 = -4 + 8 + 80 = 84$. The vertex is the point (.5, 84)

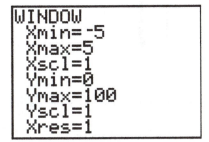 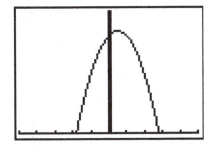 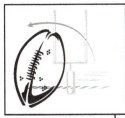

I said at the very beginning of this Booklet "…anything that you throw, toss, chuck, shoot, lob, eject, or otherwise hurl into the air will chart a parabola if you graph the distance above the ground versus time." This is always true. Sometimes, however, the actual *path* of an object can also be graphed as a parabola, but not always. For example, if you throw your pen straight up into the air and catch it, the *path* of the pen was straight up and then straight down…not a parabola. However, Brett Favre, when he is attempting a 75 yard touchdown pass will throw the football in the path of a parabola.

> Sometimes the *path* of an object ejected in the air looks like a parabola, sometimes it does not.
>
> However, the graph of the objects distance above the ground versus time is <u>always</u> a parabola.

The quadratic function $s(t) = -(1/2)gt^2 + V_0t + H_0$ is a special quadratic function related to physics. In fact, the example above is one of these. This function represents the object's distance above the ground (s) as a function of time (t), given an initial velocity (V_0) and an initial position or height (H_0). The constant g is the acceleration of gravity on earth, which is 32 ft/s^2 or about 9.8 m/s^2. The reason this is negative is because gravity acts downward. If the object is thrown downward, then the initial velocity would also be negative.

Example: Write the quadratic function representing the distance above the ground versus time for a College Algebra textbook that is thrown upwards, off a 50 ft. roof with an initial velocity of 12 ft/s^2.

$$s(t) = -(1/2)gt^2 + V_0t + H_0 = -(1/2)\cdot 32t^2 + 12t + 50 = \underline{-16t^2 + 12t + 50}$$

Example: Write the quadratic function representing the distance above the ground versus time for a surface-to-air missile launched from the deck of an aircraft carrier 245 feet above sea level with an initial velocity of 120 ft/s^2. Graph the parabola using an appropriate window and find when the missile reaches its highest point and the height at that point.

$$s(t) = -(1/2)gt^2 + V_0t + H_0 = -(1/2)\cdot 32t^2 + 120t + 245 = \underline{-16t^2 + 120t + 245}$$

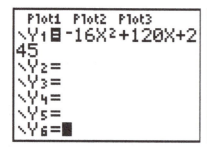 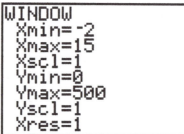 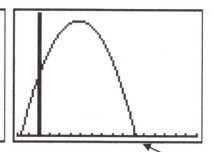

The missile is at its highest point at the vertex of the parabola. The vertex is t = -b/2a which = -120/(2*-16) = 120/32 = 3.75 seconds.

The height is the y-coordinate of the vertex, or s(3.75) = -16(3.75)^2 + 120(3.75) + 245 = 470 feet.

The point (t, s(t)) = (3.75, 470) shows that after 3.75 seconds of flight, the missile is 470 feet above sea level.

From the graph above, we can estimate that the missile takes about 9 seconds to hit its target, but how would we solve for that *algebraically*? Good question! And as soon you go through this next practice pause, I will show you!

VIP – VERY IMPORTANT POINT

The parabola above does not give us the distance the missile traveled, just its height above the ground at any time, t. We know that the missile hit the ground after about 9 seconds of flight, but we do not know how far it traveled.

Professor's Practice Pause

(1) Given the function $g(x) = 0.25x^2 + 3$, describe as much as possible about the parabola in writing and then use your TI to graph the parabola in an appropriate window. Be sure to consider the roles of a, b, and c.

Finding the vertex using the TI-83 is fairly simple. If the parabola opens up, we use the `minimum` feature, and if it opens down, we use the `maximum` feature. Using the missile equation and graph above, press **2nd TRACE 4**: (which is `maximum`). ◄────

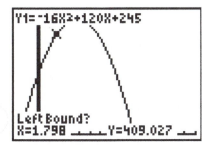

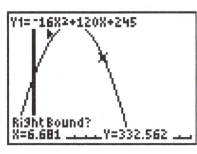

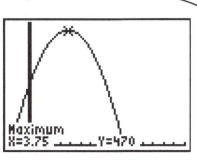

Use the arrow key to move the blinking cursor anywhere to the left of the	Now use the arrow key to move the blinking cursor anywhere to the right of the vertex and press ENTER.	Press ENTER for `Guess?` and we should see the same answer that we calculated above...

(2) Calculate the vertex of the parabola graphed using $y = .75x^2 + 6x - 3$ and then verify your answer by using the TI and the `minimum` feature (just choose **3** instead of **4** after **2nd TRACE**).

Associated End of Booklet exercise is 11

Solving Quadratic Equations

We will learn how to solve quadratic equations using both algebraic techniques and technology. Below is a little outline of the methods we will cover:

> ## Methods of solving quadratic equations
>
> A. Algebraic techniques
>
>   I. Factoring with Zero Product Principle (ZPP)
>   a. Common factor
>   b. "Unfoiling"
>   *i.* Perfect square
>   *ii.* Difference of squares
>   c. Completing the square
>
>   II. The Quadratic formula
>
> B. TI-83 using the Intersect Method

The solutions to the quadratic equation $ax^2 + bx + c = 0$ have several names:

> a. Solutions
> b. x-intercepts
> c. x-axis crossings
> d. Horizontal intercepts
> e. Zeros
> f. Roots

> For the most part, these are all synonyms and can be used interchangeably. You will notice a slight difference in meaning once we solve quadratic equations with *complex* roots. In this text, I will most often use the term "roots."

Why use the word *root* to describe where the parabola crosses the x-axis? Consider a tree…the tree is the parabola, the x-axis is the ground, so where does the tree intersect the ground?…at the roots. Where does the parabola (tree) cross the x-axis (ground)?…at the roots.

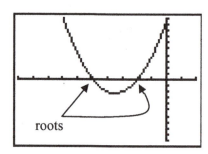

roots

…analogous to…

This graph above is also the same as the equation $ax^2 + bx + c = 0$.

"parabola"

"crosses"

x – axis

> Asking "when does the equation $ax^2 + bx + c$ equal 0?"
> is the same as asking
> "Where does the parabola cross the x – axis?"

Why factor?

There are many, many types of factoring. Some math texts will spend so much time emphasizing all the different flavors and varieties that students just end up more confused – and that's an important *factor* (ha!). I will only emphasize a few techniques so you have the basics and an understanding of *why* we factor in the first place.

Many students ask me, "What is factoring all about?" or "Why do we factor?" The answer is very simple – and here is the reason…we factor to turn several terms added and subtracted together into several terms multiplied together. Dah – dah!

Ok…apparently the importance of that last sentence did not wend its way into the connections of your hippocampus. I'll say it again…

We factor to turn several terms added and subtracted together into several terms multiplied together. Why is that important? Because once we have $a \cdot b = 0$ (instead of $a + b = 0$) we can use the Zero Product Principle (ZPP) and continue solving.

Aight! Get the beat-box goin'. "You down with ZPP, yeah you know me! You down with ZPP, yeah you know me!" *Naughty by Nature* presented us with that lovely Old-Skool rap song back in the day…you're down with a discount!

ZPP does not mean Other People's Property (as in the song), it means Zero Product Principle.

The ZPP is an incredibly simple, yet powerful principle in mathematics. It goes like this…
If you have two numbers, terms, items (whatever) and you multiply them together and the answer is 0, then either one or the other or both *must* equal 0.

In math, it looks like this…

$$\text{If } a \cdot b = 0, \text{ then either}$$
$$a = 0\ldots\text{or}$$
$$b = 0\ldots\text{or}$$
$$a = b = 0$$

It doesn't just apply to *two* items; it can be any finite number of items.

If $A \cdot B \cdot C \cdot D \cdot E = 0$, then either $A = 0$ or $B = 0$ or $C = 0$ or $D = 0$ or $E = 0$ or any combination of them equals 0 or all of them equal 0.

Now we can start putting these together…

Common Factor

Factoring is dividing. When your 10th grade substitute math teacher told you to "factor out" something, what she really was saying was "divide it out."

Examples:

Original expression	greatest common factor	factored expression
$3x^2 - 4x$	x	$x(3x - 4)$
$-2x + 8$	2	$2(-x + 4)$
$6y - 4y^2$	2y	$2y(3 - 2y)$
$3p^4 + 6p^3 + 9p^2$	$3p^2$	$3p^2(p^2 + 2p + 3)$

So what do I mean by "dividing out?" Let's look at the first row in more detail.

$$3x^2 - 4x = \frac{x}{x}(3x^2 - 4x) = x\left(\frac{3x^2}{x} - \frac{4x}{x}\right) = x(3x - 4)$$

Notice that when the x is factored out, you can find out what remains in the parentheses by dividing each term by that same x.

If you are having difficulty understanding factoring, it may be good to do it this way

Let's expand our table a bit by applying the ZPP in order to set up the solution of our equations (our expressions set equal to 0).

Original equation	Factored equation	Using the ZPP If $a·b = 0$, then $a = 0$ & $b = 0$
$3x^2 - 4x = 0$	$x(3x - 4) = 0$	$x = 0$ & $3x - 4 = 0$
$-2x + 8 = 0$	$2(-x + 4) = 0$	$2 = 0$ & $-x + 4 = 0$
$6y - 4y^2 = 0$	$2y(3 - 2y) = 0$	$2y = 0$ & $3 - 2y = 0$
$3p^4 + 6p^3 + 9p^2 = 0$	$3p^2(p^2 + 2p + 3) = 0$	$3p^2 = 0$ & $p^2 + 2p + 3 = 0$

Terms added or subtracted	→	Terms multiplied together	→	Each term set equal to 0

Example: Finish row 1 in the table above.

$3x^2 - 4x = 0$ becomes $x(3x - 4) = 0$ becomes $x = 0$ and $3x - 4 = 0$

$$\underline{+4 \quad +4}$$
$$3x = 4, \text{ so } x = 4/3$$

We have two answers that *satisfy* the equation…$x = 0$ and $x = 4/3$

Example: Finish row 2 in the table above.

-2x + 8 = 0 becomes 2(-x + 4) = 0 becomes 2 = 0 and –x + 4 = 0

Clearly, 2 is not 0, so we throw that statement out. The only solution is x = 4

Exercise: You finish row 3 in the table above….

Un-foiling

Also known as "factoring a trinomial into two binomials"…do you see why I stuck with un-foiling? It's the opposite of foiling. There are many different techniques for this factoring procedure, but I am going to show you one that always works.

Foiling is the same as *expanding*, while un-foiling is the same as factoring (or condensing).

We will only consider trinomials in which the coefficient of the square term is 1. For anything else, we have the quadratic formula. So here's our problem…

Factor $x^2 + 12x + 32$ into $(x + P)(x + Q)$…..how do we find P and Q?

Step 1: Start by writing all pairs of numbers that multiply to give the *constant* term, +32

(1, 32) & (-1, -32)
(2, 16) & (-2, -16)
(4, 8) & (-4, -8)

Step 2: But only one of these 6 pairs add to give the middle term (+12)…..(4, 8)

P   Q

So, $x^2 + 12x + 32 = $ **(x + 4) (x + 8)**

Example: Factor $x^2 - 11x - 12 = 0$, use the ZPP and solve for x.

Step 1: Start by writing all pairs of numbers that multiply to give -12

(-6, 2) & (6, -2)
(-12, 1) & (12, -1)
(-4, 3) & (4, -3)

Step 2: Only one of these pairs add to give the middle term of -11…(-12, 1)

So, $x^2 - 11x - 12 = $ **(x – 12)(x + 1)**

Since (x – 12)(x + 1) = 0, we can set each term equal to 0 individually

$$x - 12 = 0, \text{ means } \underline{x = 12}$$
$$x + 1 = 0, \text{ means } \underline{x = -1}$$

Example: Factor $x^2 - 3x + 18 = 0$, use the ZPP and solve for x.

Step 1: Write product pairs for 18: (18, 1), (-18, -1), (3, 6), (-3, -6), (9, 2), (-9, -2)

sums: 19   -19   9   -9   11   -11

Step 2: Uh Oh! None of these pairs add to give -3. That simply means that this quadratic is not factorable using this method. That's ok…we will try something else later using this same quadratic.

Another **GOAL** of factoring is to try the easiest method first. If that fails, work yourself up to the most difficult method, which is the quadratic formula. First see if there is a common factor, then look at un-foiling, and finally try the quadratic formula. I haven't really mentioned completing the square because that technique simply leads us to the quadratic formula….as we will see.

Two special cases of Un-foiling

Difference of squares and Perfect squares make life a little easier if you can recognize them. A difference of squares is exactly that…the subtraction of two, square numbers. For example: $x^2 - 25$, $x^2 - 9$, $b^2 - 100$, $x^2 - y^2$, $t^2 - 64$, etc…

Remember from the Booklet on Complex numbers (I know, I know, that was a long time ago!!). We showed that a complex number multiplied by its complex conjugate is always a real number…this is an example of *difference of squares*.

We had $(a + bi)(a - bi) = a^2 + abi - abi - (bi)^2$, which reduced to $a^2 - (bi)^2$. If we think backwards (how do you do that anyway?) we see that $a^2 - (bi)^2$ comes from $(a + bi)(a - bi)$. Now apply that to some of our differences of squares examples from above…

$$x^2 - 25 = (x + 5)(x - 5) \qquad b^2 - 100 = (b + 10)(b - 10) \qquad x^2 - y^2 = (x + y)(x - y)$$

In *general* the solutions to $x^2 - a^2$ are $x = +a$ and $x = -a$

In block diagrams, it looks like this…

Question??….is $x^2 - 15$ a difference of squares? We know that x^2 is a square, but what about 15? What number, multiplied by itself, gives 15?…How about $\sqrt{15}$.

$$x^2 - 15 = (x + \sqrt{15})(x - \sqrt{15})$$

Again, thinking backwards we can see what a *perfect square* is all about. We know that $(x + 3)^2$ = $(x + 3)(x + 3) = x^2 + 6x + 9$. In this case, the product pair is the same number repeated $(3, 3)$

$$x^2 + 10x + 25 = (x + 5)(x + 5) = (x + 5)^2 \qquad x^2 - 4x + 4 = (x - 2)(x - 2) = (x - 2)^2$$

In general, this looks like $x^2 + (2a)x + a^2 = (x + a)(x + a) = (x + a)^2$ and the solutions are $x = -a$ and $x = -a$, or $x = +a$ and $x = +a$…the same solution twice. A parabola represented by a perfect square quadratic only has one (repeated) solution.

Professor's Practice Pause

(1) Explain why we factor.

(2) Explain the Zero Product Principle (ZPP).

(3) Factor the following expressions. Use the common factor technique, the un-foiling technique, or both, to factor (condense) the expression as much as possible.

(a) $z^2 - 2z - 15$

(b) $x^2 - 3x + 2$

(c) $2x^2 - 8x$

(d) $5p^2 + 35p$

(e) $y^2 - 8y$

(f) $x^2 - 12x + 36$

(g) $t^2 - 81$

(h) $\mu^2 - \varphi^2$

(i) $3x^2 - 15x - 42$

(j) $4z^2 - 10$

(4) Using the factored expression in (a) – (j) above, set them equal to 0 and solve using the ZPP.

(a)

(b)

(c)

(d)

(e)

(f)

(g)

(h)

(i)

(j)

Associated End of Booklet exercises are 12 and 13

Completing the Square (CTS)

I am showing you this final factoring technique not so much so you can solve quadratic equations by applying it, but because it is used to derive the quadratic formula.

Completing the square is performed by following a sequence of steps...

Step 1: Make sure the leading coefficient (a) is 1. If not, divide the equation through by a.

Step 2: Calculate one-half of the middle coefficient and square it...that is $(b/2)^2$.

Step 3: Add and Subtract $(b/2)^2$ to the left side of the equation.

(This is a trick used in math quite often to change the *form* of an expression without changing its *value*. If we both add and subtract the same amount, what have we really added?........0.)

Step 4: Now group together the square term, the linear term, and the added $(b/2)^2$ term, so that it becomes possible to un-foil those three as a perfect square. The perfect square will always be $(x + b/2)^2$.

Step 5: Solve the equation by moving the remaining constants to the right side and taking the square root of both sides.

Example: Solve the equation $x^2 + 6x - 11 = 0$ by completing the square.

Before doing anything, let's check the easier factoring methods first.

~ Is there a common factor? No!
~ Is it a perfect square or difference of squares? No!
~ Can we un-foil the quadratic expression? Since the only product pairs are (-11, 1) and (11, -1), and neither of those add up to 6...No! Ok, we move to CTS.

Step 1: a = 1...good.
Step 2: $(b/2)^2 = (6/2)^2 = 3^2 = 9$
Step 3: We both add and subtract 9 to the left-hand side...

$$x^2 + 6x +9 - 9 - 11 = 0$$

We can do this to only one side of the equation because we have added 0.

Step 4: $(x^2 + 6x +9) - 9 - 11 = 0$
$(x + 3)^2 - 9 - 11 = 0$

Step 5: solve the equation...
$(x + 3)^2 - 9 - 11 = 0$.....move the -20 to the other side
$(x + 3)^2 = 20$.............now take the square root of both sides
$x + 3 = \pm\sqrt{20}$... → ...$x = \sqrt{20} - 3$ and $-\sqrt{20} - 3$...or...x = 1.47 & -7.47

Let's check this graphically using the TI…

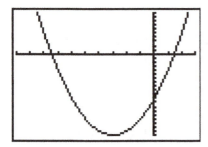

By inspection, the roots look to be at about -7.5 and 1.5, which confirms our work above. Also notice that the graphed parabola intercepts the y-axis at (0, -11) which we expect from the equation. The parabola opens up and the line of symmetry is not the y-axis, which we also expect from the equation (because a > 0, and b ≠ 0).

Derivation of the Quadratic Formula

Warning – big word alert

Perhaps some of you remember this from high school?…the dreaded quadratic formula (or QF, as I like to call it). You may have even been taught an auditory *mnemonic* device to the tune of "Pop goes the weasel" by that lovely substitute, Ms. Dimplecakes. Not *Pop goes the Weasel* by 3rd Bass…that was something completely different! I mean the nursery rhyme…you know… "All around the mulberry bush" and all that jazz.

The great thing about the QF is that it solves _every_ quadratic equation!

In fact, it even solves quadratics where the roots are complex numbers. ((Now might be a good time to go back to the *Booklet on Complex Numbers* and at least read the Topical Summary to refresh your memory))…and while you're at it, go grab a bowl of Fruit Loops or something – that's what I'm about to do…

Where does the quadratic equation come from?

We know that $x = -\dfrac{b - \sqrt{b^2 - 4ac}}{2a}$ and $x = -\dfrac{b + \sqrt{b^2 - 4ac}}{2a}$

But where does that equation come from? It comes from CTS on the general form of the quadratic, $ax^2 + bx + c = 0$

We start as follows……… $ax^2 + bx + c = 0$

(Remember, this represents every possible quadratic because it is in symbolic form)

Step1: We divide by a to get "1" in front of the x^2 term…. $x^2 + \dfrac{bx}{a} + \dfrac{c}{a} = 0$

(This step is optional now) We subtract the c/a from both sides… $x^2 + \dfrac{bx}{a} = -\dfrac{c}{a}$

Step 2: Calculate one-half of the middle term squared.

$(1/2)*(b/a) = b/(2a). \quad [b/(2a)]^2 = b^2/4a^2.$

Step 3: Add and subtract $b^2/4a^2$ to the left hand side.

$$x^2 + \frac{b\,x}{a} + \underbrace{\frac{b^2}{4\,a^2} - \frac{b^2}{4\,a^2}}_{\text{added \& subtracted}} = -\frac{c}{a}$$

(another optional step…move the $-b^2/4a^2$ over the right side to leave only the perfect square trinomial on the left side)

$$x^2 + \frac{b\,x}{a} + \frac{b^2}{4\,a^2} = \frac{b^2}{4\,a^2} - \frac{c}{a}$$

Step 4: The perfect square is always " x plus one-half the middle term" quantity squared. In this case we have…

$$\left(x + \frac{b}{2\,a}\right)^2 = \frac{b^2}{4\,a^2} - \frac{c}{a}$$

Step 5: Solve the equation…this one is a little tricky! First, get a common denominator on the right hand side. The LCD would be $4a^2$, so we multiply the c/a by (4a)/(4a) and combine, which gives…

$$\left(x + \frac{b}{2\,a}\right)^2 = \frac{b^2 - 4\,a\,c}{4\,a^2}$$

Now we take the square root of both sides…

$$x + \frac{b}{2\,a} = \sqrt{\frac{b^2 - 4\,a\,c}{4\,a^2}}$$

** Remember that $\sqrt{\dfrac{x}{y}} = \dfrac{\sqrt{x}}{\sqrt{y}}$ **

The bottom of the right hand side can be simplified to 2a…

$$x + \frac{b}{2\,a} = \frac{\pm\sqrt{b^2 - 4\,a\,c}}{2\,a}$$

Note that when we take the square root of the right hand side there are 2 answers, positive and negative…that's why we always have the + answer and the – answer)

Finally, we subtract b/(2a) from both sides giving the form of the quadratic formula…

$$x = -\frac{b}{2\,a} + \frac{\sqrt{b^2 - 4\,a\,c}}{2\,a} \quad \text{and} \quad x = -\frac{b}{2\,a} - \frac{\sqrt{b^2 - 4\,a\,c}}{2\,a}$$

Why did we break up the fractions in the QF? Look at the next page…

A graphical representation of the Quadratic Formula

If we let A = $-\dfrac{b}{2a}$ and B = $\dfrac{\sqrt{b^2-4ac}}{2a}$ then the QF boils down to just x = A ± B, right?

Do you recognize what A represents?...think...that's the x-coordinate of the vertex of a parabola – which is also the axis of symmetry. Since B is just a number, what we have is "x = the middle of the parabola plus and minus some number (distance)." What this is telling us is that the two real roots of a parabola are the same distance (B) away from the middle. With that information, we can get a graphical representation of the QF.

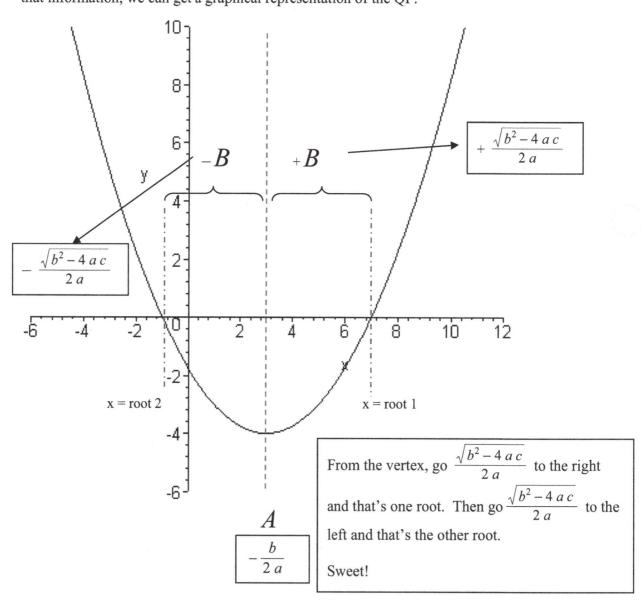

$-\dfrac{\sqrt{b^2-4ac}}{2a}$

$+\dfrac{\sqrt{b^2-4ac}}{2a}$

$-B$ $+B$

x = root 2

x = root 1

A

$-\dfrac{b}{2a}$

From the vertex, go $\dfrac{\sqrt{b^2-4ac}}{2a}$ to the right and that's one root. Then go $\dfrac{\sqrt{b^2-4ac}}{2a}$ to the left and that's the other root.

Sweet!

Example: Use the QF to find the roots of $x^2 - 2x - 6 = 0$.

It's a good idea to first list a, b, and c. a = 1, b = -2, and c = -6. Now substitute these values into the QF…

$$x = \frac{-(-2) \pm \sqrt{((-2)^2 - 4(1)(-6))}}{2(1)} = \frac{2 \pm \sqrt{(4 + 24)}}{2} = \frac{1 \pm \sqrt{28}}{2} = 1 \pm \sqrt{7}$$

The roots are $x_1 = 1 + \sqrt{7}$ and $x_2 = 1 - \sqrt{7}$.

Example: Use the QF to find the roots of an un-factorable quadratic, namely $-2x^2 - 5x + 1 = 0$.

List a, b, and c. a = -2, b = -5, and c = 1. Now simply "plug and chug" as it were.

$$x = \frac{-(-5) \pm \sqrt{((-5)^2 - 4(-2)(1))}}{2(-2)} = \frac{-5 \pm \sqrt{(25 + 8)}}{4} = -1.25 \pm 1.436$$

The roots are $x_1 = -1.25 + 1.436 = \underline{0.186}$ and $x_2 = -1.25 - 1.436 = -\underline{2.686}$. It was easier to switch to decimal, so that's what I did.

Our solutions are the points (0.186, 0) and (-2.686, 0)

Professor's Practice Pause

(1) Use CTS to factor the quadratic expression on the left hand side and solve the equation $x^2 + 10x - 3 = 0$. Express your answers in exact (radical) form and also in decimal form.

(2) Write the factoring form that you would use to solve each of the following equations and explain why you chose that particular method. You do not actually need to solve…
    (a) $x^2 - 9 = 0$ (b) $x^2 + 3x - 10 = 0$

    (c) $3x^2 + .86x - 1.24 = 0$ (d) $4x^2 + 2x = 0$

(3) Use the QF to find the roots of the equation $0 = .5x^2 - 6.5x + 12.75$

Associated End of Booklet exercises are 14 and 15

One more example of using the QF...

Solve $x^2 + 3x + 4 = 0$

First, try something easier…is there a common factor?...No. Is the trinomial a perfect square or difference of squares?...No. Can the trinomial be un-foiled?...No. Ok, good, so let's use the QF.

Listing the coefficients, we have a = 1, b = 3 and c = 4.

$$x = \frac{-(3)}{2(1)} \pm \frac{\sqrt{(3)^2 - 4(1)(4)}}{2(1)} = \frac{-3}{2} \pm \frac{\sqrt{9 - 16}}{2} = \frac{-3}{2} \pm \frac{\sqrt{(-7)}}{2} \ldots??$$

Uh Oh! Houston we have a problem! We know that we cannot have negative values under the square root, right? Our answers are not real numbers, they are *complex* numbers (Dorothy we're not in Kansas anymore!!)

Let's continue the problem using our Memory Break info….

$$x = \frac{-3}{2} \pm \frac{\sqrt{(-7)}}{2} = \frac{-3}{2} + \frac{\sqrt{7}\,i}{2}$$

Memory Break
Remember from the Booklet on Complex Numbers that $\sqrt{(-x)} = \sqrt{x}\cdot i$, where i is the imaginary unit.

The complex roots of this equation are $x_1 = (-3 + \sqrt{7}\,i)/2$ and $x_2 = (-3 - \sqrt{7}\,i)/2$. Remember we stated that complex numbers always come in conjugate pairs…$(a + bi)$ and $(a - bi)$.

In decimal form, our answers are $x_1 = -1.5 + 1.323i$ and $x_2 = -1.5 - 1.323i$.

Where do these answers live?...on the complex plane. The graph of our parabola does not cross the x-axis, so there are no real roots. The roots are complex! Stare at the graphs on the top of the next page and see if they make any sense…

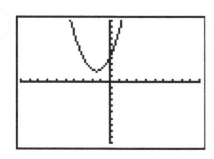

← No x-axis crossings mean no real roots.

Complex conjugate roots → are found on the complex plane.

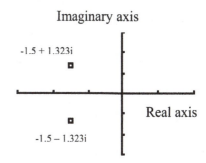

Imaginary axis

-1.5 + 1.323i

Real axis

-1.5 – 1.323i

Example of complex roots: Use the QF to solve the following equation. Express the roots in exact form and in decimal form. $x^2 + 8 = 0$

Listing the coefficients we get a = 1, b = 0 (there is no x term), and c = 8

$$x = \frac{0}{2(1)} \pm \frac{\sqrt{0^2 - 4(1)(8)}}{2(1)} = \pm \frac{\sqrt{-32}}{2} = \pm \frac{\sqrt{32}\,i}{2} = \pm \frac{\sqrt{16}\sqrt{2}\,i}{2} = \pm \frac{4\sqrt{2}\,i}{2} = \pm 2\sqrt{2}\,i$$

In decimal form, we have $0 + 2.828i$ and $0 - 2.828i$, or simply $\pm\, 2.828i$.

Notice we could have also (and more simply) solved this equation algebraically…

$x^2 + 8 = 0$…means…$x^2 = -8$…taking the square root of both sides gives…$x = \pm\, \sqrt{-8}$
So, $x = \pm\, \sqrt{8}\,i = \pm\, 2\sqrt{2}\,i$. The same answer, just quicker!

NOTE: Whenever a quadratic equation has **no** *linear* term, the second method above should probably be used…it tends to be quicker!

The Discriminant

Not discriminate…discrim*inant*. A funny word, but a useful concept. The discriminant of the QF is the term $b^2 - 4ac$. Sometimes, math people use the "delta" symbol from the linear slope equation to represent the discriminant. $\Delta = b^2 - 4ac$.

Valuable information can be gained knowing the sign of Δ, as shown in the table below:

Sign of Δ	Type of Roots	Description of Parabola
$\Delta < 0$	Two complex conjugate roots	As seen in the above examples, this parabola does not cross the x-axis.
$\Delta > 0$	Two distinct real roots	This parabola crosses the x-axis in two different locations.
$\Delta = 0$	One real repeated root	This parabola has one real root repeated, as in the solution to a perfect square.

The Match Game

Match the value of Δ with the letter of the appropriate equation and the letter of the appropriate graph. Note that the equations given are not those of the graphs shown.

Sign of Δ	Letter of equation	Letter of graph
Δ < 0		
Δ = 0		
Δ > 0		

Equations:

(a) $x^2 - 2x + 1$

(b) $2x^2 + 4x - 12$

(c) $3x^2 - x + 5$

(a)	(b)	(c)

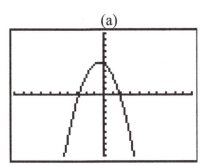
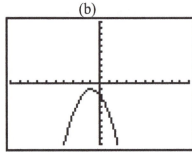
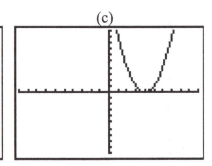

And the last method of solving quadratics is...

…using technology. As Kip would say from Napolean Dynamite, "But I still love my technology…" We've already talked about the 6 magic button (also called the intersect method). Using it to find roots of quadratics is very simple: Put the expression in Y_1, put 0 in Y_2 and press **2nd, TRACE, 5, ENTER, ENTER, ENTER**. As an example, I will show you the different screens for solving $-2x^2 - 5x + 1 = 0$. We did this example just before using the QF.

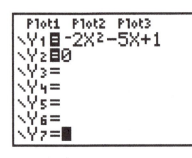

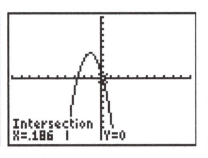

The first answer is x = .186, which agrees with our work by hand.

In order to find the 2nd root, we press the 6 buttons over again, but we need to do something before we press the last ENTER…

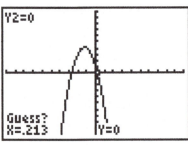

Before you press ENTER for the 3rd time, you have to move the little cursor *closer* to the other root using the arrow keys. It doesn't have to be

exactly on it………→

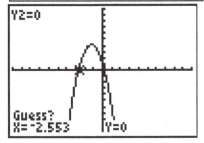

Now press ENTER for the 3rd (last) time…and viola!

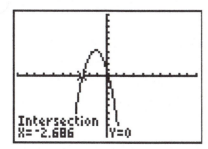

Intersection
X=-2.686   Y=0

So basically you are using the Intersect method twice…once for each root. The only catch is that before you press ENTER for the last time when finding the 2nd root, use the arrow keys to move the cursor over closer to that 2nd root.

NOTE: The TI method can only be used for finding real roots. Since complex roots do not cross the x-axis, the Intersect method will not work. The only way to find complex roots is to use the quadratic formula (or factoring in some cases), or…you can use the TI-83 program that I wrote and will share with you at the end of this Booklet (if you're allowed, that is…)

Group Pause

In groups of 2 or 3, solve each equation algebraically and check your answers using the Intersect method when appropriate.

(a) $4x^2 + 12x - 16 = 0$

(b) $2.5x^2 - 8.09x = 4.65$

(c) $x^2 + 13 = 0$

Associated End of Booklet exercises are 16 – 21

Real World Application...

Another characteristic of the parabola is the *focus*. The focus is a point that has a very interesting property when it comes to "focusing" waves such as light, sound, etc. Any wave approaching the parabola parallel to the line of symmetry will be reflected directly toward the focus. Similarly, any wave emanating from the focus will be reflected by the parabola parallel to the line of symmetry.

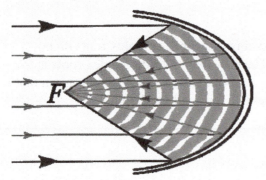

The picture to the left is what we would call a *parabolic reflector*. Satellite dishes, communication satellites, magnifying glasses, telescopes, and many other devices employ this property of reflection through the focus or focal point, F.

Source: Wikipedia

According to legend, our good friend Archimedes created parabolic mirrors to burn the decks of the Roman ships when they attacked Syracuse (not NY). You may have also unwittingly used this feature in your destruction of ants and other small creatures under the power of your magnifying glass when you were but a child (*unwittingly* being the operative word).

The focus of a parabola is a point given by the following coordinates…

$$F = \left(-\frac{b}{2a}, \frac{1 - b^2 + 4ac}{4a} \right)$$

Notice that the x-coordinate of the focus is the same as the vertex. This is because the focus lies on the axis of symmetry. The focus *is not on* the parabola, which makes the derivation of the coordinates slightly beyond the scope of this text.

Example: Given the quadratic function, $f(x) = .5x^2 - 2x - 4$, find the focal point of the parabola generated by $f(x)$.

We first identify a, b, and c…a = .5, b = -2, and c = -4.

The x-coordinate = -b/(2a) = -(-2)/(2*.5) = 2.

The y-coordinate = $(1 - b^2 + 4ac)/(4a)$ =
$(1 - (-2)^2 + 4(.5)(-4))/(4*.5) = -11/2 = -5.5$

The focus is the point (2, -5.5).
By comparison, the vertex is (2, -6)

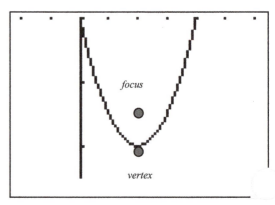

Modeling Quadratic functions (QuadReg)

Now would be a good time to go back to the *Booklet on Linear Functions* and brush-up on the procedure for finding the **regression** equation (equation of best fit). The only difference is that now we will be finding the quadratic equation of best fit, and the only change in the procedure is to choose 5:QuadReg, instead of 4:LinReg(ax+b), as shown below.

```
EDIT CALC TESTS
1:1-Var Stats
2:2-Var Stats
3:Med-Med
4:LinReg(ax+b)
5:QuadReg
6:CubicReg
7↓QuartReg
```

Example: The table below shows the estimated population of the United States for a given year. This data comes from the US Bureau of the Census.

Year	Population (millions)
1900	75.996
1910	91.972
1920	105.711
1930	122.775
1940	131.669
1950	150.697
1960	179.323
1970	203.185
1980	226.546
1990	248.710
2000	281.421

TASK 1: Enter the data into your TI, using years as the input variable and population as the output.

Remember our trick of making 1900 be year 0, 1910 be year 10, and so on. So, really our input variable is number of years from 1900.

Create a scatter plot of your data in an appropriate window. It should look like the one below…

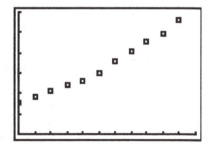

TASK 2: Find the quadratic regression equation of best fit. As a quick reminder, you will be pressing **STAT**, right arrow over to **CALC** and press **5:**. If you used Lists 1 and 2, you would then press **2nd 1 comma 2nd 2**. You should see…

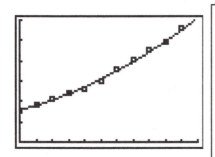

The regression equation is y or f(x) = .009x^2 + 1.077x + 78.32.

Notice that there is no r-value. That's ok, we can calculate it by taking the square root of R^2.

The correlation coefficient (r) is .999. Very nice match!

TASK 3: Just to verify, enter the equation in Y_1 and graph it over the scatter plot. Do you remember the quick way to do that??

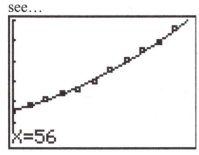

Wait a minute there mister author!! You said that population growth was best modeled by exponential functions! What gives?

Yes, I did. Take a minute and find the *exponential regression equation* for the same data. What do you notice about the r-value?

r = _____

TASK 4: based on your equation, what should the US population be in 2010? (Are we interpolating or extrapolating?). Algebraically, we are finding f(110) = .009(110)^2 + 1.077(110) + 78.32 = 305.69. Our estimate of the population in 2010 is almost 306 million. (As of 1/1/08, the population is 303.14 million…looks like we're on track with this math model).

We could have also used the Table method or **2nd**, **Trace**, **1:value** to find the same answer.

TASK 5: What about the population in 1956? That's not given in the table. In this case, we would need to find f(56) = .009(56)^2 + 1.077(56) + 78.32, or using **2nd**, **Trace**, **1:value**, we see…

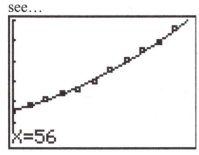

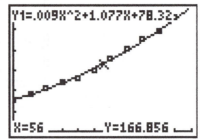

According to the US Census Bureau, the actual population in 1956 was 168.9 million.

Group Pause

Dr. Raymond Q. Bixby, the great piscatorially-preoccupied percussionist (he plays Bass and catches Bass), has created a new boomerang. He climbs 35 feet to the top of the new dorm, *Frisbie* Hall, and throws the boomerang into the air. Unfortunately, the boomerang does not come back to him, but just floats gently to the ground.

His trusty assistant, (you), records the height of the boomerang at a given time along its path of flight. The data is below.

Time (seconds)	0	8	14	21
Height (feet)	35	46	38	10

1. Algebraically find the expression that describes the boomerangs height as a function of time. *[[HINT: start with $f(x) = ax^2 + bx + c$. You know c (or do you?) Pick two data points and set up a system of linear equations to find a and b]]*

2. Use the QuadReg function of your TI to find the quadratic of best fit. Do your answers to 1 and 2 match?

3. Find the maximum height of the boomerang.

4. Find the number of seconds it takes for the boomerang to hit the ground.

Associated End of Booklet exercises are 22 – 28

A note about Domain and Range of quadratics...

Finding the domain and range of quadratics (polynomials in general) is pretty simple, so we won't beat this to death!

Any real number substituted for the input variable will yield a real number for the output, so it should be clear that the domain of every quadratic (and every polynomial as we will see) is all real numbers. Look at the parabola examples below...

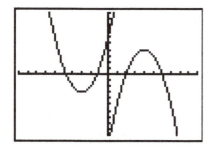

Notice that there are no leftmost or rightmost values. Since the parabola continues forever in both directions, every x (input) value has a corresponding point on the curve.

$D_{quadratic}$ = all real numbers = $(-\infty, \infty)$

However, it is also clear from the graphs above that the output values (the graph, functional, or curve values) start at a certain point, right? And that certain point is the vertex in every case. If the parabola opens up (a > 0), the vertex is the minimum value and if the parabola opens down (a < 0), the vertex is the maximum. With that in mind, we have these definitions for the range of a quadratic:

If a > 0, $R_{quadratic}$ is from the vertex to infinity...[y-coordinate of vertex, ∞)

If a < 0, $R_{quadratic}$ is from negative infinity to the vertex...$(-\infty$, y-coordinate of vertex]

Topical Summary of Quadratic Functions

Take about 20 – 30 minutes and create your own summary of this chapter. Go back, review, and write below all of the main points, concepts, equations, relationships, etc. This will help lock the concepts in your brain now, and provide an excellent study guide for any assessments later.

End of Booklet Exercises (What goes up, must come down!)

(1) State whether each function below is a quadratic function. If not, explain why.

(a) $f(x) = 3x^2 - 5x + 2$

(b) $g(x) = 3x^2 - 5x$

(c) $h(x) = -3x^2$

(d) $p(x) = 3x$

(e) $g(x) = 3x^2 - 5x + x^3$

(f) $f(x) = 1 - x^{-2}$

(2) Algebraically calculate the vertex for each parabola listed below.

(a) $y = 6x^2$

(b) $y = -.5x^2 + 4x - 9$

(c) $f(x) = x - x^2$

(d) $y = \dfrac{7x^2}{4} + \dfrac{11x}{4} + \dfrac{2}{3}$

(e) $g(x) = 10x^2 + 4$

(f) $y = \pi x^2 + \mathbf{e}x + \alpha$

(3) For all the quadratics in question 2, go back and write the equation for the line of symmetry.
    (a)   (b)   (c)

    (d)   (e)   (f)

(4) The vertex of a parabola is the point (2/3, -7/3). The y-intercept of the parabola is (0, -1).
Find the equation of this parabola, if you also know that $b = -(4/3)a$.

(5) The vertex of a parabola is the point (-.25, -7.25). The y-intercept of the parabola is (0, -7).
Find the equation of this parabola, if you also know that $a = 2b$.

(6) Explain the roles of the constants a, b, and c in the general quadratic $f(x) = ax^2 + bx + c$.

(7) The leading coefficient of a parabola is negative. What information does this provide about the graph of the parabola?

(8) Write an example of a quadratic function, f(x), in which the parabola opens up and the y-intercept is the point (0, 12).

(9) Write an example of a quadratic function, f(x), in which the parabola opens down and the vertex is the origin.

(10) Without the use of your calculator, write the number of the parabola next to the matching equation.

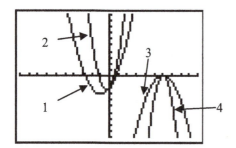

$f(x) = -x^2 + 12x - 36$	_____
$f(x) = x^2 + 2x - 2$	_____
$f(x) = -4x^2 + 48x - 144$	_____
$f(x) = 3x^2 + 2x - 2$	_____

(11) Verify your answers using the minimum or maximum feature on the calculator for all of the vertices you calculated in question (2). If any answer from question 2 does not match the TI answer, go back and re—check your work.

(a) _____   (b) _____   (c) _____

(d) _____   (e) _____   (f) _____

(12) Factor the quadratics below using any of the techniques you learned in class (common factor, un-foiling, difference of squares, perfect square).

(a) $x^2 - 81$ (b) $x^2 + 8x + 16$ (c) $x^2 + 2x - 15$

(d) $4x^2 - 100$ (e) $2x^2 + 16x + 32$ (f) $x^2 + 3x - 4$

(g) $4x^2 + 8x$ (h) $2x^2 + 22x + 36$ (i) $5x^2 - 5$

(j) $x^2 - 14$ (k) $x^2 + 12x + 36$ (l) $Ax^2 - B$

(13) For each of the factored quadratics in question (12), set them equal to 0 and use the Zero Product Principle (ZPP) to find the solutions

   (a)   (b)   (c)

   (d)   (e)   (f)

   (g)   (h)   (i)

   (j)   (k)   (l)

(14) Write the quadratic formula in the space below. After you have written it, go back in the Booklet and make sure it is correct ☺

(15) Solve the equations below by completing the square.

(a) $x^2 - 4x - 3 = 0$

(b) $x^2 + 6x - 11 = 0$

(c) $-2x^2 + 5x + 13 = 0$

(d) $ax^2 - 4x + 10 = 0$

(16) Write the quadratic formula in the space below. After you have written it, go back in the Booklet and make sure it is correct ☺

(17) Use the quadratic formula to solve the following equations.

(a) $5x^2 - 4x - 3 = 0$

(b) $-2x^2 + 7.5x - 1 = 0$

(c) $4x^2 - 9 = 0$

(d) $x^2 + 3.75x - 1.5 = 0$

(e) $-2x^2 + 7x = 3 - 4x^2$

(f) $\dfrac{2+x}{3} = \dfrac{5}{x}$

(g) $x^2 - 3x + 4 = 6 - 3x^2 + 11x$

(h) $\alpha x^2 + \beta x + \gamma = 0$

(18) For problems (a) – (g) in question (17), use the Intersect method on the TI to verify your answers. If the TI answers do not match yours, go back and re-check your work. Write the TI solutions on the spaces below.

(a) _____

(b) _____

(c) _____

(d) _____

(e) _____

(f) _____

(g) _____

(19) **Going back to the Golden Ratio...**

From page 6...Cut a length into two pieces, a larger and smaller. Call the larger section a and the smaller section b.

a   b

a + b

If the ratio of the whole length to the larger is the same as the ratio of the larger to the smaller, and if both of those lengths equal φ, then the lengths are said to be sectioned according to the golden ratio.

$$\frac{a+b}{a} = \frac{a}{b} = \varphi .$$ where $$\varphi = \frac{1+\sqrt{5}}{2} \approx 1.61803\,39887\ldots$$

But where does the 1.618 come from? The derivation goes like this...From the second equation ($\frac{a}{b} = \phi$) we get $a = b\,\phi$. Substituting in the first equation gives us $\frac{b\phi + b}{b\phi} = \frac{b\phi}{b}$, which

simplifies to $1 + \frac{1}{\phi} = \phi$. Finding a common denominator (ϕ) on the left hand side gives

$\frac{\phi+1}{\phi} = \phi$. Multiplying both sides by ϕ gives $\phi^2 = \phi + 1$. Setting the equation equal to 0

gives $\phi^2 - \phi - 1 = 0$. Solve this equation. Remember, since ϕ represents a ratio, only the positive answer has meaning.

(20) Write the quadratic formula in the space below. After you have written it, go back in the Booklet and make sure it is correct ☺

(21) Use the quadratic formula to solve the equations below. **Note**: Since none of the parabolas intersect the x-axis, all of your solutions will involve *complex conjugate pairs*.

(a) $5x^2 - 4x + 3 = 0$

(b) $2x^2 + 0.5x + 1 = 0$

(c) $4x^2 + 9 = 0$

(d) $x^2 + 3.75x + 4.5 = 0$

(e) $2x^2 + 7x + 15 = 0$

(f) $\dfrac{2-x}{3} = \dfrac{5}{x}$

(22) A rocket launcher mounted on the deck of a US naval destroyer fires a rocket at an approaching enemy fighter plane. The deck of the destroyer is 25' above sea level. The table below represents the height of the rocket above sea level as a function of time.

Time (seconds)	Height above Sea level (feet)
0	30
2	45
3	67
5	125
7	180
8	248
12	420
15	496

(a) Use your TI to find the quadratic regression equation of best fit. Write the equation and the r-value below.

(b) Use your equation to predict the height of the missile above sea level after 20 seconds (extrapolation).

(c) Use your equation to predict the height of the missile above sea level after 9 seconds (interpolation).

(23) The product of two numbers is -63. The sum of the two numbers is -2. Write two separate equations and then combine them to form a quadratic equation. Use the quadratic equation to find the two numbers.

(24) The y-intercept of a parabola is (0, 4). The parabola also contains the points (-2, -18) and (3, 22). Find the equation of the parabola by setting up a system of linear equations and solving for a and b, similar to the last **Group Pause** in this Booklet.

(25) The product of two numbers is -20.482. The sum of the numbers is 5.91. Find the two numbers.

(26) The product of two consecutive, odd integers is 3135. Find the two integers.

27) The product of two consecutive, positive, even integers is 624. Find the two integers.

(28) One number is 10 more than 4 times the other number. Their product is 266. Find the two numbers. Both numbers are positive.

Booklet on the Higher Order Polynomials

What is a polynomial?

We used the term polynomial in the Booklet on Quadratic functions, and we also kind of have an idea of what a polynomial is from the table at the beginning of that Booklet. The general form of a polynomial looks like this:

$$P(x) = a_m x^n + a_{m-1} x^{n-1} + a_{m-2} x^{n-2} + a_{m-3} x^{n-3} + \ldots + a_2 x^2 + a_1 x + a_0$$

The a's represent the coefficients of each term, where a $\in$ **R**.
x is the (single) variable
m is the index of the coefficient
n is the exponent shown in decreasing order

What makes a polynomial a polynomial is the restriction on the exponent, *n*. *n* must be a positive integer ($n \in$ **N**)

Examples of polynomials	Examples of non-polynomials
(a) All quadratics	(a) $x^{-2} + 4x - 8$ (negative exponent)
(b) $-5x^3 - 3x^2 + 2x - 9.67$ (c) $10x^3 + 4x - 1$ (d) x^3 $\}$ cubics	(b) $5x^{1/4} - 6x^2 + 3x - 7$ (fractional exponents)
(e) $.5x^7 + \pi x^6 + 8x - 2$	(c) $x^\pi - 5x + 1$ (irrational exponent)
(f) $2x^{100} + x^{50} - 5$	

The **degree** (or **order**) of a polynomial is the highest exponent. All quadratics are degree 2; Cubics are degree 3; Quartics are degree 4, Quintics are degree 5, and so on. The degree of the polynomial in example (e) above is 7, while the degree of example (f) is 100.

Creation Break...

Write any polynomial function of degree 5, with variable *s*, that has 4 terms.

P(s) = _____

Professor's Practice Pause

For the expressions listed in (a) – (f), state whether each is a polynomial or not. If it is a polynomial, state the order. If not, explain why not.

(a) $1 - x^2 + x - 5\,x^4$

(b) $x + .75\,x^3 - \dfrac{1}{x}$

(c) $x^{17} - 1$

(d) $\pi\,x + \mathbf{e}\,x^2 - .983\,x^{12}$

(e) $-3\,x^{(-1)} + x - x^2 + 4$

(f) $x^n + x^{(n-1)} - 3\,x^{(n-3)} + 5\,x^{(n-5)}$, where $n = 3$

Associated End of Booklet exercises are 1 and 2

Solving Higher Order Polynomial equations

The fundamental Theorem of Algebra is sometimes stated as: *Every non-zero single-variable polynomial, with complex coefficients, has exactly as many complex roots as its degree, if each root is counted up to its multiplicity.*

Huh?

Basically, it says that any polynomial with one variable (like *x*) has the same number of roots (solutions) as the highest exponent. Sometimes the roots are repeated (like in quadratics). If that happens we say that a root has multiplicity 2, or 3, or 4, depending upon how many times it shows up as a solution.

Degree of a polynomial = Number of roots

For example, the quadratic equation $x^2 + 6x + 9 = 0$ can be re-written as $(x + 3)(x + 3) = 0$, which means that the roots are $x = -3$ and $x = -3$. We can simply say that the roots are $x = -3$, *multiplicity* 2.

Our Goal

Our goal in solving higher order polynomials is very simple. We use a factoring technique (or several techniques) to simplify the polynomial until we've reached one with a degree of 2 (quadratic). Then we use the QF.

The factoring techniques for solving higher order polynomials are primarily the same as for solving quadratics (factoring, difference of squares, quadratic forms, etc), it's just that there are *more* solutions (more than 2).

There are two new techniques introduced in this Booklet…Quadratic forms and Polynomial Long Division. We'll see about those a bit later.

Example 1: Solve $x^4 - 1 = 0$

We can use difference of squares to get $(x^2 - 1)(x^2 + 1) = 0$
Then use the ZPP (remember that?) to separate and solve each one…

$(x^2 + 1) = 0$ $(x^2 - 1) = 0$
$x^2 = -1$ $x^2 = 1$
$x = \pm \sqrt{-1}$ $x = \pm 1$
$x = \pm i$

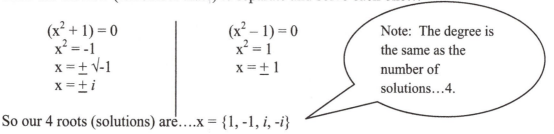

Note: The degree is the same as the number of solutions…4.

So our 4 roots (solutions) are….$x = \{1, -1, i, -i\}$

Notice that the graph of $x^4 - 1$ looks like a parabola (but a little flatter at the bottom). You can see the 2 real roots (at +1 and -1) on the TI-83, and since you know the total number of solutions to be 4 (highest exponent), you then can determine that there are also 2 complex roots…but that doesn't tell you what they are.

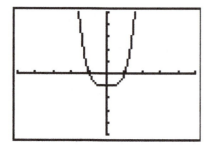

This is a powerful discovery tool! Graphing a polynomial shows you how many real roots there are by counting the number of x-axis crossings. Once you know that, you can subtract the number of real roots from the degree to get the number of complex roots.

Number of complex roots = Degree – Number of real roots

Example 2: Solve $x^5 - 5x^3 - 6x = 0$

Factor out an x from each term to get $x (x^4 - 5x^2 - 6) = 0$
When you use the ZPP to set each factor equal to 0, you get <u>x = 0</u> for the first root

Now, how do we solve $x^4 - 5x^2 - 6 = 0$…?

We can un-foil this to get $(x^2 - 6)(x^2 + 1) = 0$….and now (again), separate….

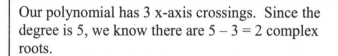

$$(x^2 + 1) = 0 \qquad\qquad (x^2 - 6) = 0$$
$$x^2 = -1 \qquad\qquad\qquad x^2 = 6$$
$$x = \pm \sqrt{-1} \qquad\qquad x = \pm \sqrt{6}$$
$$x = \pm i$$

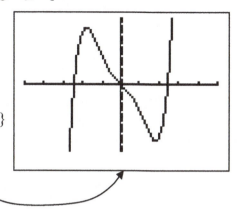

So our 5 roots (solutions) are….$x = \{0, \sqrt{6}, -\sqrt{6}, i, -i\}$

Our polynomial has 3 x-axis crossings. Since the degree is 5, we know there are $5 - 3 = 2$ complex roots.

{{ ***Reminder***: complex roots come in *conjugate pairs*! }}

Example 3: Solve $x^3 - 4x^2 - 11x + 30 = 0$

Since we won't be able to factor this polynomial, let's look at using the Intersect Method to solve for the roots.

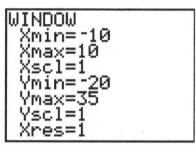
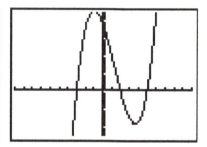

Remember our trick…before pressing ENTER for the 3rd time when solving for the second and third roots, use the arrow keys to move the cursor close to that root. I just happened to solve for the middle one first, then the left root and finally the right one.

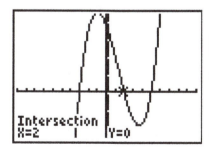
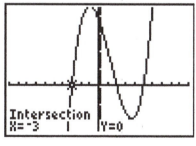
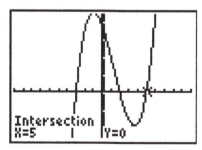

Our solutions to $x^3 - 4x^2 - 11x + 30 = 0$ are $x = \{-3, 2, 5\}$

Re-writing in factored form...

Since we know the roots, we can re-write $x^3 - 4x^2 - 11x + 30$ in factored form as follows:

$$x^3 - 4x^2 - 11x + 30 = (x + 3)(x - 2)(x + 5).....\text{why that?}$$

Remember from the ZPP, when we set $(x - a) = 0$ and solve we get $x = a$. So, if a is a root of $P(x) = 0$, then $(x - a)$ is a factor of $P(x)$, and vice-versa.

> If *a* is root of some polynomial equation $P(x) = 0$, then $(x - a)$ is a factor of the polynomial expression.
>
> In other words, **(x – root) = factor**

Going back to **Example 1**, $x^4 - 1$ can be re-written in factored form as follows...

The roots were $x = \{1, -1, i, -i\}$, so we have $x^4 - 1 = (x - 1)(x - (-1))(x - i)(x - (-i))$

$= (x - 1)(x + 1)(x - i)(x + i)$

Group Pause

In groups of 2 or 3, solve the following equations. Use factoring, the TI-83, the quadratic formula, or any combination of these techniques to solve.

(a) $x^4 - 25 = 0$

(b) $x^3 - 4x^2 - 21x = 0$

(c) $x^3 - 6x^2 + 13x = 0$

(d) $x^4 + 2x^2 - 3 = 0$

Associated End of Booklet exercise is 3

Example 4:   Solve $x^3 - 1 = 0$

If you graph $x^3 - 1$, you will notice that it crosses that x-axis at $x = 1$ (the only real root). That means there are 2 complex roots also.

The Intersect Method cannot solve for the remaining roots because they are complex and we can't use difference of squares since the exponent is *odd*. How do we solve this? Ahhhh…we can use something from way, way back in your Elementary School education…long division.

Ahhh…Elementary School! Memories of snack pack puddings in Scooby Doo, semi-rusted, metal lunch boxes…the screams of little pony-tailed girls when you pull their hair or put a dead worm on their desk…the odor of chalk mixed with whatever chemicals they used to clean the floors. Everything was big back then; desks, teachers, the playground, the vice-principal's paddle (ouch!!). But no other educational memory has woven its way permanently into the synapse network of the brain more so than long division. ((If this is absolutely *not* true for you, take a quick peek back at the *Booklet on Real Numbers* for a refresher on long division))

Polynomial Long Division

Back to $x^3 - 1 = 0$. Since we know that $x = 1$ is a *root*, we also know that $(x - 1)$ is a *factor*, right?

(x – root) = factor.

That means $(x - 1) \cdot$ (some other factor) $= x^3 - 1$……what is the other factor?

Divide both sides by $(x - 1)$ to get … (other factor) $= (x^3 - 1) / (x - 1)$

Set up the right hand side like a long-division problem…$(x - 1)$ *goes into* $(x^3 - 1)$

You may remember this as "Guzinta"

There's no such thing as Guzinta. It's a figment of Mr. Krackenfuss' imagination!

Which is like *Gesundheit*, but not quite…

Mr. Krackenfuss is a figment of *my* imagination!

We'll set up the polynomial division problem right next to a numerical division problem so we can see the great similarity between the two.

$7\overline{)164}$ Write the problem in LD form

$x - 1\overline{)x^3 - 1}$ Since x multiplied by x^2 gives x^3, then x goes into x^3, x^2 times.

$\begin{array}{r} 2 \\ 7\overline{)164} \end{array}$ How many times does 7 go into 16?…2 times.

$\begin{array}{r} x^2 \\ x - 1\overline{)x^3 - 1} \end{array}$ Notice we only use the highest term of each to do the division.

$$\begin{array}{r} 2 \\ 7\overline{)164} \\ 14 \end{array}$$

Multiply. 2 times 7 is 14 and we write that under the 16.

$$\begin{array}{r} x^2 \\ x-1\overline{)x^3-1} \\ x^3-x^2 \end{array}$$

$x^2(x-1) = x^3 - x^2$ and we write that under the $x^3 - 1$.

When we multiply back, we multiply the entire divisor by x^2.

$$\begin{array}{r} 2 \\ 7\overline{)164} \\ -14 \\ \hline 2 \end{array}$$

Subtract. We subtract 14 from 16 and get 2

$$\begin{array}{r} x^2 \\ x-1\overline{)x^3-1} \\ -(x^3-x^2) \\ \hline x^2-1 \end{array}$$

We must subtract the whole expression, hence the parentheses. $(x^3 - 1) - (x^3 - x^2) =$ $x^3 - 1 - x^3 + x^2 = x^2 - 1$

$$\begin{array}{r} 2 \\ 7\overline{)164} \\ -14 \\ \hline 24 \end{array}$$

Bring down. We bring down the 4

$$\begin{array}{r} x^2 \\ x-1\overline{)x^3-1} \\ -(x^3-x^2) \\ \hline x^2-1 \end{array}$$

There are no other terms so we don't bring down anything.

$$\begin{array}{r} 23 \\ 7\overline{)164} \\ -14 \\ \hline 24 \end{array}$$

Start over again…
Divide. How many times does 7 go into 24?...3 times

$$\begin{array}{r} x^2+x \\ x-1\overline{)x^3-1} \\ -(x^3-x^2) \\ \hline x^2-1 \end{array}$$

Looking at highest terms, x goes into x^2, x times.

$$\begin{array}{r} 23 \\ 7\overline{)164} \\ -14 \\ \hline 24 \\ 21 \end{array}$$

Multiply. 3 times 7 is 21 so we write that under the 24

$$\begin{array}{r} x^2+x \\ x-1\overline{)x^3-1} \\ -(x^3-x^2) \\ \hline x^2-1 \\ x^2-x \end{array}$$

$x(x-1) = x^2 - x.$

$$\begin{array}{r} 23 \\ 7\overline{)164} \\ -14 \\ \hline 24 \\ -21 \\ \hline 3 \end{array}$$

Subtract. $24 - 21 = 3$

$$\begin{array}{r} x^2+x \\ x-1\overline{)x^3-1} \\ -(x^3-x^2) \\ \hline x^2-1 \\ -(x^2-x) \\ \hline x-1 \end{array}$$

$x^2 - 1 - (x^2 - x) =$

$x^2 - 1 - x^2 + x = x - 1$

Since 7 does not "go into" 3, we are done.
The answer is 23 *and* 3/7.
164/7 = 23 *and* 3/7
The answer is **not** 23 r3..!!!

x – 1 divides x – 1 exactly 1 time, so our answer is $x^2 + x + 1$
$(x^3 - 1)/(x - 1) = x^2 + x + 1$
$\underline{x^2 + x + 1 \text{ is the other factor.}}$

So $x^3 - 1$ can be written as $(x - 1)(x^2 + x + 1)$

$(x - 1)(x^2 + x + 1) = 0$....now use the ZPP

$x - 1 = 0$, so $x = 1$ (which we already knew from the graph).....and....

$x^2 + x + 1 = 0$ (we have to use the QF because there are 2 complex roots)
$(a = b = c = 1)$

$$x = \frac{-1/2 \pm \sqrt{(1 - 4(1)(1))}}{2} = \frac{-1/2 \pm \sqrt{(-3)}}{2} = -1/2 \pm \sqrt{(3)}/2 \, i$$

So our 3 roots (solutions) are....$x = \{1, \ -\dfrac{1}{2} + \dfrac{\sqrt{3}\,i}{2}, \ -\dfrac{1}{2} - \dfrac{\sqrt{3}\,i}{2}\}$

Mini Rant and Rave Pause

Let me rant again for just a second. The reason I don't like the "r" notation for remainder is because it can lead to misunderstanding, and the wrong solution. Let me show you how…

Using the "r" notation, $100/99 = 1$ r1.

$1000/999 = 1$ r1

$10000/9999 = 1$ r1

This would lead you to believe that all three answers are the same. But they are not!

$100/99 = 1$ *and* $1/99 = 1.01$

$1000/999 = 1$ *and* $1/999 = 1.001$

$10000/9999 = 1$ *and* $1/9999 = 1.0001$

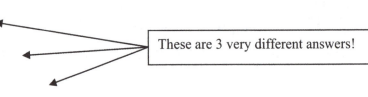

These are 3 very different answers!

Thank you! I feel better now ☺

Professor's Practice Pause

Use the TI and/or factoring with polynomial long division to solve for the roots of the following equation...

$$x^3 - 4x^2 + 6x - 9 = 0$$

Associated End of Booklet exercises are 4 and 5

Modeling Higher Order Polynomial functions (CubicReg, QuartReg)

Let's go back to the rocket launcher problem from the Booklet on Quadratic Functions. Given the data below, we calculated the quadratic regression equation to be $f(x) = .88x^2 + 20.62x + 10$, which generated an r-value of $r = 0.993$. Very good!

A rocket launcher mounted on the deck of a US naval destroyer fires a rocket at an approaching enemy fighter plane. The deck of the destroyer is 25' above sea level. The table below represents the height of the rocket above sea level as a function of time.

Time (seconds)	Height above Sea level (feet)
0	30
2	45
3	67
5	125
7	180
8	248
12	420
15	496

We can see if perhaps a higher order polynomial model will give us the same (or even slightly better) results.

Use your TI to generate 3rd and 4th order equations for the data above, and show the corresponding r-values.

CubicReg: _____

QuartReg: _____

What does this little exercise lead you to believe about the relationship between the r-value and the order of polynomial? Write your hypothesis below...

Hypothesis about r-value and order of polynomial regression...

Quadratic Forms

The QF is useful for solving quadratic equations in which x equals two answers. But we can also use the QF to find answers to any equation, as long as the form looks like…

$$(\text{something})^2 + \text{something} + \text{number} = 0$$

We call this a *quadratic form*, because it fits the form of the quadratic. For example, the equation $3x^4 + 2x^2 - 10 = 0$ is a quadratic form. How do I know? I ask myself 'what is being squared in the first term?' The answer is x^2. x^2 is being squared, so I can re-write the equation to look like…

$$3(x^2)^2 + 2(x^2) - 10 = 0$$

We go about solving the same way. First list a, b, and c…a = 3, b = 2, and c = -10.

Now, instead of x = …, we have $x^2 = \dfrac{-(2)}{2(3)} \pm \dfrac{\sqrt{2^2 - 4(3)(-10)}}{2(3)} = \dfrac{-1}{3} \pm \dfrac{\sqrt{31}}{3} = 1.523$ *and* -2.189.

We're not done. Our equations look like $x^2 = 1.523$ and $x^2 = -2.189$

If we take the square roots of all 4 sides, we get the following answers:

$$x_1 = 1.234, x_2 = -1.234, x_3 = 1.479i, x_4 = -1.479i$$

We notice some things…
 (1) The complex roots come in conjugate pairs. In this case, so do the real roots.
 (2) There are 4 roots, which makes sense because the highest term is x^4.

Example 2: Use quadratic forms to solve for the roots of $2e^{2x} - 3e^x - 5 = 0$.

It may take some coaxing to think about what it is being squared in this example, so to help you along, I'll re-write the equation as follows: $2(e^x)^2 - 3(e^x) - 5 = 0$. We can now see that e^x is being squared, so instead of e^x we write **u** or some other letter.

Let $u = e^x$. With that substitution the equation becomes $2u^2 - 3u - 5 = 0$. Using the quadratic formula we end up with u = -1 and u = 2.5.

However, $u = e^x$. So we have two separate equations to solve…

$e^x = -1$ | $e^x = 2.5$

Since e^x can never be < 0, this equation does not produce any solutions.

$\ln(e^x) = \ln(2.5)$
$x = \ln(2.5) \sim 0.916$

In this case, there is only one solution.

Let's verify this solution graphically...

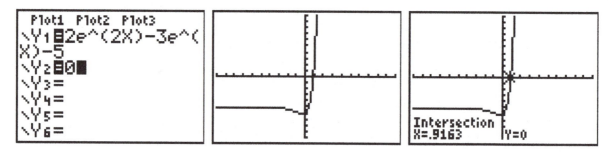

Professor's Practice Pause...

Use the *quadratic form* technique to solve the following equation. Verify your answer using the Intersect method on the TI-83. {Hint: You may also need to incorporate some other techniques as well…}

$$2(\ln(x))^2 - 4\ln(x) - 16 = 0$$

Associated End of Booklet exercise is 6

Factors to roots / roots to factors

We know so far that the *factor = x – root*. This allows us to work backwards, if you will. So far we have been finding the roots given the equation, as such…

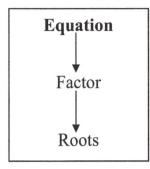

But we should be able to work the other way as well…like this…

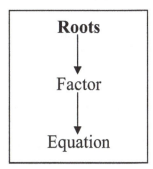

Starting with the roots, we can then write the factors and the equation by multiplying the factors together.

EXAMPLE 1: Write the quadratic equation that has roots 5 and -3.

We start by writing the factors $(x – 5)$ and $(x – (-3))$, or $(x – 5)$ and $(x + 3)$

We FOIL the factors to get the equation… $(x – 5)(x + 3) = \underline{x^2 – 2x – 15}$

EXAMPLE 2: Write the cubic equation that has roots 2, 0 and -4.

We start by writing the factors $(x – 2)$, $(x – 0)$ or just x, and $(x + 4)$

We FOIL the factors to get the equation… $x(x – 2)(x + 4) = \underline{x^3 + 2x^2 – 8x}$

Well that seems easy enough, eh?? But what if the roots are complex….

EXAMPLE 3: Write the quadratic equation that has roots $2 – i$ and $2 + i$.

Always start by writing the factors first…$(x – (2 – i))$ and $(x – (2 + i))$

Now we FOIL the factors…

$(x – (2 – i))\cdot(x – (2 + i)) = x^2 – x(2 + i) – x(2 – i) + (2 – i)(2 + i)$

Expanding, we get…$x^2 – 2x – xi – 2x + xi + (2 – i)(2 + i)$

Notice the *i* terms cancel

From the Booklet on Complex Numbers…the product of a complex number and its conjugate is always $a^2 + b^2$. In this

That leaves us with $\underline{x^2 – 4x + 5}$. You can verify this result by solving this quadratic using the QF.

EXAMPLE 4: Write the cubic equation that has roots $4 - 3i$ and -6.

A warning message should be going off in your head right now. If not, you may need to reboot your brain quickly...that usually works for me.

I asked you to find the *cubic*, but I only gave you $\underline{2}$ roots. We know that all cubics have 3 roots, so what's up? Which one is missing? Well, if we remember that complex roots come in complex conjugate pairs, that tells us that $4 + 3i$ must be another root (along with $4 - 3i$).

So our factors are $(x + 6) \cdot (x - (4 - 3i)) (x - (4 + 3i))$

It's easier to first FOIL the complex pair and then multiply by $(x + 6)$ afterwards. I leave this exercise as your Practice Pause for the day!

Professor's Practice Pause...

Find the equation of the cubic with roots $4 - 3i$, $4 + 3i$ and -6.

Associated End of Booklet exercises is 7

Pascal's Triangle and the Binomial Expansion

This is an optional concept for this text since it is usually studied in Statistics, Pre-calculus, or another math class. However, since there are applications to factoring and FOIL-ing, your professor may want to cover this in class. You might be able to bribe her by bringing assorted baked goods into the classroom...you never know!

Oh well...if that failed then here goes!!

You know how to FOIL $(x + 2)^2$.... $= (x + 2)(x + 2) = x^2 + 4x + 4$

Well, what if I said to expand $(x + 2)^3$? That would be $(x + 2) (x + 2) (x + 2)$, which is our previous result $(x^2 + 4x + 4)$ times another factor of $(x + 2) = x^3 + 6x^2 + 12x + 8$ ...(you can also use the box method for that problem)

Let's be a little more generic....$(x + a)^2 = x^2 + ax + a^2$

What about $(x + a)^3$? That would be $x^3 + 3 x^2 a + 3 x a^2 + a^3$

What about $(x + a)^{10}$? You wouldn't want to multiply all of these would you?
$(x + a) (x + a) (x + a) (x + a) (x + a) (x + a) (x + a) (x + a) (x + a) (x + a) =$ _____

I don't think so. Even as a form of punishment, this exercise would be cruel and inhuman.

Pascal's triangle is a simple way to find the coefficients of the "binomial expansion." The general binomial expansion is $(x + a)^n$. So how do we construct Pascal's triangle?

We build it as follows....

$$
\begin{array}{ccc}
 & 1 & \\
1 & & 1
\end{array}
$$

start by writing "1"
then write 1 and another 1

Each row has one more element than the one above and the ends are always 1's. The way you find the middle is by adding up the elements above to the left and right....so the next row would be 1 2 1, because $1 + 1 = 2$

$$
\begin{array}{ccccc}
 & & 1 & & \\
 & 1 & & 1 & \\
1 & & 2 & & 1
\end{array}
$$

The next would be   1   3   3   1, because $1 + 2 = 3$ and $2 + 1 = 3$

$$
\begin{array}{ccccccc}
 & & & 1 & & & \\
 & & 1 & & 1 & & \\
 & 1 & & 2 & & 1 & \\
1 & & 3 & & 3 & & 1
\end{array}
$$

Then so on…..

$$\begin{array}{ccccccc}
 & & & 1 & & & \\
 & & 1 & & 1 & & \\
 & 1 & & 2 & & 1 & \\
1 & & 3 & & 3 & & 1 \\
\end{array}$$

```
                      1
                   1      1
                1      2      1
             1      3      3      1
          1      4      6      4      1
       1      5     10     10      5      1
    1      6     15     20     15      6      1.........and so on
```

Each **row** corresponds to an increasing exponent (**n**) in $(x + a)^n$.

```
n = 0                           1
n = 1                        1      1
n = 2                     1      2      1
n = 3                  1      3      3      1
n = 4               1      4      6      4      1
n = 5            1      5     10     10      5      1
n = 6         1      6     15     20     15      6      1……….
```

Remember these numbers are the **coefficients** of the terms. Compare above with $(x + a)^3$. That would be $x^3 + 3\,x^2\,a + 3\,x\,a^2 + a^3$

When n= 3, we have 1   3   3   1…just like the triangle says. You will notice that the terms themselves follow a pattern also…the first term in the binomial (x) starts at x^n, and then the exponent decreases by 1 in each term until you get x^0.

The second term in the binomial (a) starts at a^0 and goes up to a^n

EXAMPLE 1: Expand $(x + y)^5$ using Pascal's triangle

The x's are written as follows…$x^5 + x^4 + x^3 + x^2 + x^1 + x^0$

And the y's are written in the opposite manner…. $y^0 + y^1 + y^2 + y^3 + y^4 + y^5$

Now put them together….

$\underline{\quad}x5y^0 + \underline{\quad}x^4y^1 + \underline{\quad}x^3y^2 + \underline{\quad}x^2y^3 + \underline{\quad}x^1y^4 + \underline{\quad}x^0y^5$

The blanks are the coefficients from Pascal's triangle for the row n=5.

$\underline{1}\,x5y^0 + \underline{5}\,x^4y^1 + \underline{10}\,x^3y^2 + \underline{10}\,x^2y^3 + \underline{5}\,x^1y^4 + \underline{1}\,x^0y^5$

Since anything to the 0 power is 1 and we don't need to show coefficients of 1, we can clean this up a bit to show that

$$(x+y)^5 = x^5 + 5x^4y + 10x^3y^2 + 10x^2y^3 + 5xy^4 + y^5$$

EXAMPLE 2: Expand $(2x-3)^6$ using Pascal's triangle

In this example, our "x" is 2x and our "y" is (-3). In other words (x + y) is (2x + (-3)).

The x-terms are $(2x)^6 + (2x)^5 + (2x)^4 + (2x)^3 + (2x)^2 + (2x)^1 + (2x)^0$

The y-terms are $(-3)^0 + (-3)^1 + (-3)^2 + (-3)^3 + (-3)^4 + (-3)^5 + (-3)^6$

Let's put them together....

$$(2x)^6(-3)^0 + (2x)^5(-3)^1 + (2x)^4(-3)^2 + (2x)^3(-3)^3 + (2x)^2(-3)^4 + (2x)^1(-3)^5 + (2x)^0(-3)^6$$

Now we add the coefficients from Pascal's triangle...n = 6

$$1(2x)^6(-3)^0 + 6(2x)^5(-3)^1 + 15(2x)^4(-3)^2 + 20(2x)^3(-3)^3 + 15(2x)^2(-3)^4 + 6(2x)^1(-3)^5 + 1(2x)^0(-3)^6$$

In this example we are not quite done yet because we can simplify each term of the expression by raising both the (2x) factors and the (-3) factors to their appropriate exponents. Let's do that slowly...

$$= 1 \cdot 2^6 x^6 \cdot 1 + 6 \cdot 2^5 x^5 \cdot -3 + 15 \cdot 2^4 x^4 \cdot 9 + 20 \cdot 2^3 x^3 \cdot -27 + 15 \cdot 2^2 x^2 \cdot 81 + 6 \cdot 2x \cdot -243 + 1 \cdot 1 \cdot 729$$

$$= \mathbf{64x^6 - 576x^5 + 2160x^4 - 4320x^3 + 4860x^2 - 2916x + 729}$$

Group Pause...

Use Pascal's triangle to expand $(-x + 2y)^5$.

Associated End of Booklet exercises is 8

In-class, Group Exercise

Real World Application...volume of an open-top box

What's your favorite cereal? Mine is Reese's Puffs ® cereal. Think about how many different General Mills cereal brands there are…Now estimate how many General Mills boxes of cereal are in the largest grocery store in your town. Multiply that by the number of grocery stores in your town. How many towns are like that in your state, how many states in the country, how many countries in the world?….The number should be getting bigger, eh?

According to sources at General Mills, the company sold approximately *1 billion* boxes of cereal last year. If they lost just *one cent* per box due to waste or inefficiency, that would amount to a loss of $10,000,000. [Scientific notation: $(1 \times 10^9)(1 \times 10^{-2}) = 1 \times 10^7$]

An age old problem in manufacturing is how to maximize the volume of a product and minimize the amount of material (cardboard, for example). Since cardboard doesn't grow on trees (think about that for a second…), companies have to purchase the cardboard to sell their products (cereal in this case). If a company can put the maximum volume of Reese's Puffs in the minimum amount of cardboard, they can maximize their profit.

Let's simplify this problem a bit and look at constructing an open-top box from an 8.5" x 11" piece of cardboard. You will need the following items to complete this task:

(a) 8.5" x 11" (or larger) piece of cardboard. Paper will also work, but cardboard is best!
(b) A pair of scissors
(c) Scotch tape
(d) A ruler

To create the box, cut out a square of length x (you decide what value x is). When you do that your box will look like this…

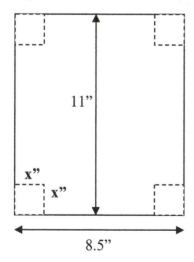

Next, fold along the cutouts to create 5 sides. After you do that, your box should look like the picture below. Keep in mind that there are only 5 sides because it is an open top.

Everyone hold up their box in class. Do you think all of the boxes have the same volume? In other words, can they hold the same amount of stuff (rice, sand, Reese's Puffs)? Some of you are saying "yes", and some of you are saying "no." You might be saying no because they all look so different – some are thin and tall, while others are wide and short. You might have said yes because you all started from the same size piece of cardboard.

The answer is no! They do not have the same volume because the volume depends upon the size of the cutout you removed from each corner. $V = f(x)$. Here's where the fun begins. We want to find the size of the cutout (x) that produces the maximum volume of the box.

(a) Find a formula for the volume of the box as a function of the size of the cutout, $V = f(x)$. Reminder: $V = L \cdot W \cdot H$

(b) Write the domain of your function? The value of x is limited, right? If $x = 5$", you wouldn't have a box anymore.

(c) Use your calculator and the domain of your function to find the x-value that gives the maximum volume. Write your answer below.

x = _____ inches

(d) What is the maximum volume? V(x) = _____ in^3.

Topical Summary of Higher Order Polynomials

Take about 20 – 30 minutes and create your own summary of this chapter. Go back, review, and write below all of the main points, concepts, equations, relationships, etc. This will help lock the concepts in your brain now, and provide an excellent study guide for any assessments later.

End of Booklet Exercises

(1) State whether or not the following expressions are polynomial expressions. If they are, state the order (or degree) of the polynomial.

(a) $\pi^z + 3z^2 - z + 1$

(b) $z^{\pi} + 3z^2 - z + 1$

(c) $y^2 - 1 + 2y$

(d) $y^2 - y^{\left(\frac{1}{2}\right)} + 2y$

(e) $x^{(-6)} - 3x^2 + 2x - 5$

(f) $x^6 - 3x^2 + 2x - 5$

(2) Write an example of a quartic polynomial with the variable s.

(3) Solve the following equations.

(a) $x^4 + 2x^3 - 35x^2 = 0$

(b) $x^3 - 10x^2 = -9x$

(c) $x^4 - 16 = 0$

(d) $x^4 - 2x^2 - 15 = 0$

(e) $x^3 - 12x^2 + 36x = 0$

(f) $-3.6x^2 + 4.87x - 10.12 = 0$

(4) Divide the following using polynomial long division.

(a) $\dfrac{x^3 - 3x^2 + 5x - 10}{x - 3}$

(b) $\dfrac{x^3 + 4x^2 - 1}{x + 2}$

(5) For the following equations, use your calculator to find a real root. Change that root to a factor and use polynomial long division to discover the remaining quadratic factor. Solve for all roots.

(a) $x^3 - 11\,x^2 + 36\,x - 26$

(b) $x^3 - \dfrac{47\,x}{4} + 17$

(6) Solve the following polynomial equations using the concept of quadratic forms.

(a) $x^4 - 2x^2 - 6 = 0$

(b) $2 \cdot 3^{2x} + 3^x - 2 = 0$

(c) $(\log_5 x)^2 + 2\log_5 x = 5$

(d) $(\ln(x))^2 - 9 = 0$

(7) In each of the problems below, find a possible polynomial equation that has the given roots.

(a) $x = \{0, 0, -1, 4\}$

(b) $x = \{3, 1 + i, 1 - i\}$

(c) $x = \{-2, -2i\}$

(d) $x = \{1, -1, i, -i\}$

(8) Expand the following using Pascal's triangle.

(a) $(a-b)^4$

(b) $(x+3)^5$

(c) $(-x+y)^4$

(d) $(2p+3q)^6$

Solutions to End of Booklet Exercises

BOOKLET 1 – Real Numbers

1) 1 2 3 4 5 6 7 8 9 10 11 12
   2 4 6 8 10 12 14 16 18 20 22 24
   3 6 9 12 15 18 21 24 27 30 33 36
   4 8 12 16 20 24 28 32 36 40 44 48
   5 10 15 20 25 30 35 40 45 50 55 60
   6 12 18 24 30 36 42 48 54 60 66 72
   …
   12 24 36 48 60 72 84 96 108 120 132 144

2) (a) -6.75, -4/3, -1, 0, e, π, 32/7, 8.5
   (b) 8.5 is real, rational
   (c) π is real, irrational
   (d) 0 is real, rational, integer
   (e) -1 is real, rational, integer
   (f) -6.75 is real, rational
   (g) 3/7 is real, rational
   (h) e is real, irrational
   (f) -4/3 is real, rational

3) No. An integer can be written as a ratio of two integers. Ex: 1 = 2/2, 4/4, etc. All ratios of two integers are rational.

4) He is wrong because 6.25/78 can be written as 625/7800, a ratio of two integers, hence rational.

5) (a) No. One-half of a number is always another number.
   (b) No. One-third of a number is always another number.
   (c) No. One-tenth of a number is always another number.
   (d) You can divide real numbers indefinitely.

6) Square roots of 2, 3, 5, 7, 11, 13, 15, 17, and 19 are irrational.

7) Answers may vary
   Q: ½, 2/3, 1.75, 0
   N: 1, 2, 3, 4
   Z: -2, 10, -50, 100
   R: π, e, 1.675, 1

8) Answers may vary…1.2, 2.5, -5.57

9) Answers may vary…2, 3, 4

10) Answers may vary…1/2, 1/3, ¼, 1/5, 1/6

11) Answers may vary…$\sqrt{2}$, $\sqrt{3}$, π, $\pi/4$, e

12) The fact that it is a perfect square means it has itself as a factor, therefore it cannot be prime.

13) It ends in 0 and it is even.

14) (a) True
     (b) False, 0 is rational
     (c) True
     (d) False, $\sqrt{2}$ is irrational, among others
     (e) True
     (f) True
     (g) False, absolute value of + is +
     (h) True
     (i) False, there is no last counting number
     (j) True
     (k) True
     (l) True
     (m) False, irrationals are a subset of reals
     (n) False, *rational* is a decimal that…
     (o) False, sum not product for perfect number
     (p) False, depends on operation
     (q) False, ! is defined as product not sum
     (r) True

15) (a) -2

(b) 2

(c) 8

(d) -8

(e) 56

(f) -4/3 = -1.333

(g) 47

(h) 121

(i) 150

(j) -5

(k) $1.03

(l) 11

(m) infinite (undefined)

(n) 0

(o) $1.82

(p) 2655 m, 8708.4 ft

16) (a) 29/21

(b) 7/45

(c) 25/4

(d) 23/24

(e) 21/4

(f) -65/6

(g) 41/24

(h) -19/6

(i) 23/66

(j) 0

17) (a) 10

(b) 10

(c) 5

(d) 5

(e) -9

(f) 1

(g) 8

(h) 11 (or 1)

(i) |G# – C#|

18)

0	0	0	0	0	0
1/5	4/15	-9/40	2/24	36/75	20/45
4/6	8/9	-36/48	20/72	72/45	40/27

18) continued..

3/16	6/24	-27/128	15/192	54/120	30/72
9/32	18/48	-81/256	45/384	162/240	10/16
1/2	2/3	-9/16	5/24	18/15	10/9
-4/10	-8/15	36/80	-4/24	-72/75	-40/45
-7/32	-14/48	63/256	-35/384	-126/240	-70/144

19) (a) 0.5 = 50%

(b) 0.666… ~ 66.6%

(c) 0.625 = 62.5%

(d) 0.777…~77.7%

(e) 0.6 = 60%

(f) 0.1666~16.6%

20) (a) 0.05 = 1/20

(b) 0.1 = 1/10

(c) 0.25 = 1/4

(d) 0.333… = 1/3

(e) 0.77 = 77/100

(f) 0.99 = 99/100

21) 25.89 inches

BOOKLET 2 – Basics of the Math Language

1) The superscript on a number or letter. It tells you to multiply what is under it by itself that many times.

2) x^3, x^3, x·x·x

3) (a) 25
   (b) 1/8
   (c) 1
   (d) 16
   (e) 16
   (f) 0
   (g) -16
   (h) 25

4) (a) 2^6
   (b) 3^6
   (c) 5^0
   (d) 4^2
   (e) $(3/5)^4$

5) (a) x^{-3}
   (b) x^3
   (c) x^{-1}
   (d) z^{-3}
   (e) y^{-1}
   (f) x^1
   (g) Q^1

6) (a) $75x^3y^6z^4$
   (b) $-48x^5y^6z^3$
   (c) $(-25/16)x^{-2}z$
   (d) rst^5
   (e) $(z^5 - z^2)/4$
   (f) 1
   (g) 1

7) $(x^n)(x^n)(x^n) = x^1$, so $3n = 1$, $n = 1/3$

8) (a) $y^{1/4}$
   (b) $(xy)^{1/4}$
   (c) $y^{2/3}$
   (d) $x^{3/5}$
   (e) $x^{4/3}y^{2/3}$
   (f) $(xyz)^{3/2}$

9) (a) $\sqrt[3]{x}$
   (b) $\sqrt{y/x}$
   (c) $\sqrt[3]{x^5}$
   (d) $1 / \sqrt{y}$
   (e) $\sqrt[3]{x^2}$
   (f) $\sqrt{x} / \sqrt{y} = \sqrt{x/y}$

10) (a) $\sqrt{2}$
    (b) $\sqrt[3]{9}$
    (c) 1/2
    (d) 5
    (e) 1
    (f) $1/(2\sqrt{2})$

11) 0.22%

12) Answers may vary

13) Answers may vary

14) Answers may vary

15) True, a*b = b*a

16) False, $a - b \neq b - a$

17) True, (a + b) + c = a + (b + c)

18) 5·7

19) $2^3 \cdot 3^2$

20) 2·3·23

21) $2^5 \cdot 3 \cdot 11$

22) 47 is prime

23)   (a) 1728
       (b) -13/49
       (c) -5.67
       (d) 6
       (e) infinite (undefined)
       (f) -6.56
       (g) 5
       (h) 1

24)   (a) 5.00×10^{-5}
       (b) 9.18×10^6
       (c) -3.29×10^1

25)   (a) -6.41×10^9
       (b) -1.99×10^{38}
       (c) 4.11×10^{19}

BOOKLET 3 – Complex Numbers

1) $i = \sqrt{-1}$

2) (a) i
    (b) $\sqrt{2}\, i$
    (c) $2i$
    (d) $9i$
    (e) $3\sqrt{5}\, i$
    (f) $6\sqrt{2}\, i$
    (g) $-2\sqrt{5}\, i$
    (h) $3 + 2\sqrt{5}\, i$
    (i) 1
    (j) xi
    (k) 1

3) (a) i. Divide 225 by 4 and let the remainder be the exponent.
    (b) 1 and 1…yes.

4) Horizontal line at $3i$

5) $2 + 4i$ and $2 - 4i$ (in the 4th quadrant)

6) $-15 + 2i$

7) (a) $6 + 4i$ product $= 52$
    (b) $-2 - 9i$ product $= 85$
    (c) The product of a complex number and its conjugate is always $a^2 + b^2$.

8) Answers may vary. Ex: $2 + 2i,\ 2 - 2i,\ 0 + \sqrt{8}\, i$

9) (a) $5 - 5i$
    (b) $-25i$
    (c) $7 - 19i$
    (d) $0.8 - 0.6i$
    (e) $6.2 - 19.47i$
    (f) $-46 - 22i$
    (g) $0.014 - 0.419i$
    (h) $2i$
    (i) $5.35 - 2.89i$

9) (j) 0

10) $\sqrt{23}\, i$

11) (a) $\sqrt{26}$
     (b) $\sqrt{109}$
     (c) $\sqrt{77.85}$
     (d) $\sqrt{(x^2 + 4)}$

BOOKLET 4 – Equations and Expressions

1) Variables: x, y, z
   Coefficients: -5, -15, -12
   Constant terms: π, 7
   Non-constant terms: $-5x^2y^3$, $-15z$, z^3, and
   $\qquad\qquad\qquad\qquad$ $-12xz$
   There are 2 expressions.

2) Answers may vary.
   One example would be $-6x^3 + 2y - x + y^2$

3) Answers may vary.
   One example would be $x + y - 8$

4) Answers may vary.
   One example would be $\sqrt{2}\, x^2y^3z^5$

5) Answers may vary.
   One example would be $3x - \pi q$

6) Answers may vary.
   One example would be $x - 1 = 0$

7) Answers may vary.
   One example would be $3x - 2x - 1 = 0$

8) Answers may vary.
   One example would be $2y + 5y = 10y$

9) After solving, we have *variable = number*.
   The variable has been isolated on one side

10) Use proper algebraic steps to arrive at
    the line where you have *variable = number*.

11) Use the order of operations (PEMDAS) to
    to reduce the expression to the fewest number of
    terms, or a single numerical value (evaluate)

12) CLT = Combine Like Terms

13) Plug the solution back into the variable in the
    original equation. If LHS = RHS, the
    solution is correct.

14) b) $x = 1/3$

15) All

16) None

17) Yes…$-1 = -1$

18) $x = 7.5$

19) $y = 0$ (answer to 17)

20) No…$3 = 1$

21) <u>Golden</u>: Whatever you do to one side of an
    equation, you must do to the other side.
    <u>Silver</u>: In order to eliminate something, you
    do the opposite operation

22) Step 2…7x is subtracted (from both sides) **S**
    Step 3…Both sides are multiplied by -1 $\quad$ **G**

23) The fraction was not broken up properly
    $(3y + 2) / 3 = 3y/3 + 2/3 = y + 2/3$

24) $x = -.08$

25) LEFT: $2x + x/2$, $x/6 - y$, $n(n + 2)(n + 4)$
    RIGHT: Quotient of twice a number and 7
    $\qquad\qquad$ Twice the sum of 5 times a number and 12
    $\qquad\qquad$ Sum of 3 consecutive odd (even) integers.

26)　(a) $7x + 3y - 5$

　　(b) $x^2 - 3x + 5y - 5$

　　(c) simplified

　　(d) 3NPQ

　　(e) $xy - 6y - 8x + 19$

　　(f) $y - x - 2$

　　(g) $2.81p + 8.9q + x + (e + \pi - 5)$

　　(h) $-2x^2 - 2x + 18$

27)　Box 1: Adding 17 to both sides

　　Box 2: Adding x to both sides

　　Box 3: Dividing by 16 on both sides

　　Box 4: $x = 20.5/16 \sim 1.28$

28) $z = -12/12.6 \sim 0.952$

29) 6 art pieces with $2 left over.

30) $(\$/g) * g = \$$

31) $v = m/s$, $a = v/s$, so $a = m/s/s = m/s * 1/s = m/s^2$

32)　(a) $p = -48$

　　(b) $p = -4/7$

　　(c) $x = -7/4$

　　(d) $x = -3/7$

　　(e) $x = 20$

　　(f) $w = -3$

　　(g) $t = -5$

　　(h) $x = 2$

　　(i) no solution

　　(j) all values of B

　　(k) $T = 52$

　　(l) $x = -8/3$

　　(m) $T = 7/20$

　　(n) $x = -24$

33)　(a) $(-1, -4)$ 　　Remember, these are

　　(b) $(10, 6)$ 　　estimates. Yours may

　　(c) $(10, -16)$ 　differ slightly.

　　(d) $(2.5, 3)$

35) $x = 5.67$

36) $x = -1$

37) no solution

OKLET 5 – The Function Concept

1) Every input value has only one output value.

2) VLT: Draw all possible vertical lines on the graph of a function. If any of the lines passes through more than one point, the graph fails the test and is not a function.

3) x: input, independent, horizontal
   y: output, dependent, vertical

4) Triangle, rectangle, hyperbola, vertical line, etc…

5) All parabolas are functions

6) Input: weight, Output: height

7) Input: hours, Output: pay

   Input: hours studied, Output: exam score

9) Input: radius, Output: area

10) Input: side length, Output: square area

11) Input: purchase price, Output: sales tax

12) Input: number, Output: square of number

13) Input: square of number, Output: number

14) Input: decimal, Output: fraction

15) Questions 6, 8, 13, and 14 are _not_ functions.

16)   (a) Yes
      (b) Yes
      (c) No

   Symbolic: $C = f(T)$
   Specific: $f(T) = 50T$

18) (a) 1, 2, 5 for example
    (b) 0, -3, 6 for example

19) (a) $f(0) = -7$
    (b) $f(-1) = -12$
    (c) $f(s) = -3s^2 + 2s - 7$
    (d) $f(.25) = -6.6875$
    (e) $f(\pi) = -30.325$
    (f) $f(\&) = -3\&^2 + 2\& - 7$
    (g) $-3red^2 + 2red - 7$
    (h) $f(1) = -8$

20) $f(4) = -6$

21) from top: G, A, D, E, B, F

22) $h(x) = (x + 2)^3 - 5$

23) $h(x) = 2(x - 1)^2 + 3$

24) $2^{(x + 2)} - 1$

25) (a) $2x^2 - 2x + 17$
    (b) $-2x^2 + 2x - 17$
    (c) No…subtraction is not commutative

26) (a) $4x^2 - 10x - 3$
    (b) $4x^2 - 10x - 3$
    (c) Yes…addition is commutative

27) $h(x) = 8x + 9$

28) $g(s) = (-1/2)s^2 + 3$

29) (a) $(-2/3)\alpha^3 + 6\alpha^2 - 13\alpha + 6$
    (b) $-2\alpha^4 + 22\alpha^3 - 59\alpha^2 + 48\alpha - 12$
    (c) $(1/3)\alpha^4 - 2\alpha^3 - (25/3)\alpha - 50$

30) $A(x) = (x^2 + x - 3)(5 + 3x)$
     $= 3x^3 + 8x^2 - 4x - 15$

31)   row 1, column 2 should be $2x^3$ (not $2x^2$)
row 2, column 3 should be $-20x$ (not $+20x$)
row 3, column 1 should be $-3x^2$ (not $-4x^2$)

32) $-3x^4 - 13x^3 + 3x^2 - 18x - 4$

33)   (a) Input: population, Output: % failures
(b) Yes, every input has exactly one output
(c) $p = G(s)$
(d) scatter plot
(e) *ii*) $f(x) = -.057x^3 + 1.2x^2 - 2.5x + 1.7$
(f) $x = 10$, so $f(10) = 39.7\%$

34)   (a) scatter plot
(b) *ii*) $f(x) = .003x^3 - 0.1x^2 + 1.38x + 31.6$
(c) $x = 50$, so $f(50) = \$225.60$

35)   (a) scatter plot
(b) *iii*) $f(x) = .275x^2 - 3.5x + 100$
(c) – (f) answers may vary

36) Yes, every input still has only one output

37)   (a) D: all real numbers, R: $y \geq 0$
(b) D: all real numbers, R: $y \geq 11$
(c) D: all real numbers, R: all real numbers
(d) D: all real numbers, R: all real numbers
(e) D: $x \neq 0$, R: $y \neq 0$
(f) D: $x \neq 0$, R: $y > 0$

38)   (a) $D = \{0, -2, 3, 5, x, \&, -\pi\}$
$R = \{0, 5, 12, y, *, e\}$
(b) $Q_{inv} = \{(0,0), (5,-2), \dots ,(*,\&), (e,-\pi)\}$
(c) $D_{Qinv} = R_Q$
$R_{Qinv} = D_Q$
(d) The answers are the opposite

39) $D = R =$ all real numbers

40) D: all real numbers, R = the y-value of the line

41) No, they are not functions.

42)   (a) -5
(b) 13
(c) $(3 - 2x)^2 - 6$
(d) $3 - 2(x^2 - 6)$

43) $(2 - (x - 1))^2$

44) One example is $B(x) = x$, $R(x) = x^2$, and $T(x) = x + 1$

45) 11

46)   (a) 0
(b) 5
(c) 3
(d) -5

47)   (a) 3
(b) 0
(c) 4. No, because $f(g(x)) \neq g(f(x))$ for all values of x

48)   (a) $5(-s^2 + 6s - 9) - 2$
(b) $\sqrt{(5s - 2)}$
(c) $-(5\sqrt{s} - 2)^2 + 6(5\sqrt{s} - 2) - 9$

49) $V = x^2y + y^2$
$SA = 4xy + x^2 + y + 2y^2/x$

50) $f(g(x)) = 3(x/3 + 5/3) - 5 =$
$x + 5 - 5 = x$
$g(f(x)) = (3x - 5)/3 + 5/3 =$
$x - 5/3 + 5/3 = x$

51)   (a) Input F, Output C
(b) $D = R =$ all real numbers
(c) $T(212) = 100$, boiling point
(d) $C_{inv}(F) = (9/5)C + 32$
(e) They meet at -40 degrees

52) (a) $(x + 3)/-10$ (b) $((x - 7)/5)^{1/2}$
(c) $x^2 + 2$ (d) $\sqrt{(x)} - 3$

OOKLET 6 – The Linear Family

1) Q1 – upper right. Q2 – upper left. Q3 – lower left
   Q4 – lower right

2)   (a) Q1
      (b) Q2
      (c) Q4
      (d) Q3
      (e) x-axis
      (f) y-axis
      (g) x-axis
      (h) y-axis
      (i) origin
      (j) Q1
      (k) Q1

3)   (a) Any positive y-value
      (b) not true
      (c) Any negative x-value
      (d) not true
      (e) $y = 0$
      (f) not true
      (g) not true
      (h) $x = y = 0$

4)   (a) 2
      (b) -2
      (c) 9/2 or 4.5
      (d) 0 (horizontal line)
      (e) infinite (vertical line)
      (f) -1.303
      (g) -.751 or -130/173
      (h) -518/15 or -34.533
      (i) $(-3 - y) / 2$
      (j) $(1 - \pi) / \pi$
      (k) $(d - b) / (c - a)$
      (l) $(\char94 - @) / (* - \$)$

5)   (a) $\sqrt{20} = 4.472$ (d) $2.5 - -12 = 14.5$
      (b) $\sqrt{405} = 20.125$ (e) $13 - -1 = 14$
      (c) $\sqrt{85} = 9.22$ (f) $\sqrt{30.629} = 5.534$

6)   (a) (1.3, 0), (0, -4)
      (b) (-1.9, 0), (0, 3.9)
      (c) (1, 0) & (3, 0), (0, 2)
      (d) none, (0, 5)

7)   (a) $m = 2, b = -7/3$
      (b) $m = -1/2, b = 7/4$
      (c) $m = -2, b = 1$
      (d) $m = -7/4, b = 5/2$
      (e) $m = 1, b = -2$
      (f) $m = 1/3, b = -2/3$

8) $m = 15, b = 58$

9) $I = f(s) = .05s + 8$

10) $k = f(O) = 7O, m = 7, b = 0$

11)   (a) $-7x + 1, -7x + 2$, etc..
      (b) $(1/7)x + 1, (1/7)x + 2$, etc..

12) $y = (3/2)x + b$ where $b \neq 2/3$

13) $y = -x + b$, where $b \neq -7$

14)   (a) $y = 2x$
      (b) $y = -2x$
      (c) $y = (9/2)x$
      (d) $y = 4$
      (e) $x = 3$
      (f) $y = -1.303x + 4.996$
      (g) $y = -.751x - 49.919$
      (h) $y = (-518/15)x + (269/15)$
         or $y = -34.533x + 17.933$

15) $y = -9x + 4$

16) $y = (1/9)x - 1$

17) $y = -x + 7$

18) y = 5x + 26

19) x = 1.25

20) Answers may vary D: 0 – 15 hours, R: 0 – 900 miles @ 60 mph.

21) Answers may vary D: 0 – 10 sec, R: 0 – 1250 ft.

22)   (a) y = -598000x + 7076000 (did you make 2004 equal to year 0?)
     (b) y = -494000x + 7318000 (did you make 2004 equal to year 0?)
     (c) Fairly close. More points for part (b)
     (d) y = 2300x + 219600
     (e) y = 10370x + 195870
     (f) Slope is far off. The points go up and down

23)   (a) scatter plot
     (b) y = 7168x + 6916, r = .973 (great match)
     (c) f(38) = 279300
     (d) Answers may vary.

24)   (a) 1938 – 1963: y = .041x + .227 (r = .992)
            1990 – 2009: y = .147x + 3.95 (r = .962)
     (b) The increases to minimum wage were less per
year in the early part of the century (4.1 cents) as opposed
to the latter part (14.7 cents)

25) Possibly completed as an in-class exercise
     (i) Scatter plot
     (ii) Yes, slightly
     (iii) Answers may vary
     (iv) Answers may vary; increase in tuition/year
     (v) Answers may vary; starting value of tuition
     (vi) Answers may vary
     (vii) Answers may vary
     (viii) Answers may vary
     (ix) y = 517.9x + 4353.5
     (x) In class
     (xi) In class
     (xii) f(26) = $17,818

BOOKLET 7 – Several Linear Families (Systems)

1) Point of intersection

2) A = C

3) -x

4) actual number = 33

5) estimates may vary…
   (a) (-1, 3)
   (b) (3, -4)
   (c) (3/2, 5)

6)   (a) (-1, 4)
   (b) (3, -4)
   (c) (0.8, 5)

7) They should be fairly close. If not go back and check.

8) 128 ounces

9) The mistake is the second equation is not in y=mx+b form. It should be –(5/4)x + 10/4. The solution is the point (.22, 2.22)

10) Intersecting – 1 solution, Equal – infinite solutions, and Parallel – 0 solutions.

11)   (a) g(x) = 4x + b, where b ≠ -7
   (b) g(x) = 4x – 7
   (c) g(x) = mx + b, where m ≠ 4

12)   (a) (-4, -5)
   (b) (4, 3)
   (c) Equal system
   (d) Parallel system

13) -40 degrees

14) (W, B) = (15, 5)

15) (B, P) = (11, 7)

16)   (a) y = (3/4)x + 3
   (b) (.48, 3.36)
   (c) 4.4

17) 57 and 93

18) x = 8 (y = -4)

19) Z = 2X – 10

20) Same answers

21)   (a) (x, y, z) = (1/3, 1/3, 1/3)
   (b) (x, y, z) = (-9, 4, 7)
   (c) (r, s, t) = (-.34, 4.81, 1.58)
   (d) (A, B, C) = (.5, .5, 0)

22) 23, 25, and 27

BOOKLET 8 – The Exponential Family

1) 10(10)(10), 5(5)(5)(5), x(x)

2) $3^4*2^4*10^4 = (3*2*10)^4 = 60^4$

3)
 (a) G
 (b) p
 (c) 1.6
 (d) 3

4) Raise 2 to the 3rd power and then multiply by 5

5) Constant slope vs. constant ratio
   No asymptote vs. asymptote (x – axis)
   Both x & y intercepts vs. only y-intercept.

6) Growing at a constant ratio of increase

7) Answers may vary

8) By the graph and by the constant ratio of successive outputs.

9) The exponents are not variable in either (a) or (b)

10) Answers may vary

11) 1^x is a horizontal line (y = 1)

12) A negative base causes the outputs to cycle back and forth between + and – values, but also between real and imaginary values.

13) The graph is reflected over the x-axis

14) Answers may vary (see question 5)

15) $f(0) = a*b^0 = a*1 = 0$

16)
 (a) not possible
 (b) Horizontal line y = 0 (x-axis)

16)
 (c) decreasing from left to right
 (d) horizontal line y = 1
 (e) increasing from left to right

17) All real numbers

18) y > 0

19) y < 0

20) All real numbers

21) A line that the curve approaches but never touches. The x-axis is an asymptote of exponential functions.

22) Exponential…population

23) Asymptote

24) As the input variable approaches infinity, the output approaches 0.

25) factor = b, rate = r when b is written as 1 – r.

26) Amount of time it takes for one-half of the radioactive material to go away.

27) 5568 years

28) calculating interest on principal more than once per year.

29) You are earning interest on the interest

30) higher interest rate (r)

31)
 (a) $5360.91…int = $360.91
 (b) $5356.13…int = $356.13
 (c) $5371.55…int = $371.55
 (d) $5373.28…int = $373.28

32) $y = 7500(1.045)^x$. P = 7500, r = 4.5%

33) (a) $17506.73
     (b) $16631.39
     (c) $500 up-front costs you $875 over 7 years

34) Linear…"million every month" constant slope.

35) $f(x) = 3*2^x$ and $g(x) = 3*3^x$

36) $f(x) = -3*2^x$ and $g(x) = -3*3^x$

37) $f(x) = -2*3^x$

38) $T(m) = -5*0.5^m$

39) $R(n) = 12*2^n$

40) 1.25

41) 1.58

42) 0.64

43) Growing because 1.82 > 1; rate = 82%

44) Decaying because 0 < .64 < 1; rate = 36%

45) One half of the radioactive compound will be gone.

46) $G(t) = 1000(1.18)^t$

47) $D(t) = 1000(0.82)^t$

48) $945*2^{(1/12)} = 1001.193$ Hz

49) (a) 125
     (b) 1/9
     (c) 16

50) $y = 150(1.25)^x$

51) (a) exp. $y = 125(.86)^x$
     (b) lin. $y = 2.45x + 85$
     (c) neither
     (d) exp. $y = -20(2.3)^x$

52) (a) D: all real number R: y > 0
     (b) D: all real number R: y > 0
     (c) D: all real number R: y < 5
     (d) D: all real number R: y < 0
     (e) D: all real number R: y > -6

53) (a) $y = 2.061(.921)^x$, r = -.982
     (b) $y = 824.4(.921)^x$, r = -.982
     (c) y2 = 400(y1)…same b and r
     (d) 1.067 ~ 16/15

54) (a) $y = 93.108(1.01)^x$, r = .947
     (b) P ~ 17 W
     (c) P ~ 34 W
     (d) Answers may vary

55) (a) 11.4
     (b) 22.8
     (c) $f(x) = 10(1/2)^{(x/11.4)}$
     (d) 100 ~ .023g: 1000 ~ 3.93E-26
     (e) No. You can always take half
         of some amount.

56) (a) $I(x) = 1.129(1.0138)^x$
     (b) $C(x) = 1.321(1.00606)^x$
     (c) 2028

57) (a) $y = 84.25(1.044)^x$, r = .886
     (b) y = 238 cents
     (c) Fairly good since the r-value is
         pretty close to 1.

BOOKLET 9 – The Logarithmic Family

1)

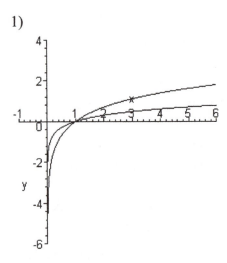

$\ln(x)$ increases faster than $\log(x)$ because the reverse is true concerning the inverse functions (10^x and e^x). Since 10^x increases faster, it is closer to the y-axis. When that is reflected over the line $y = x$, it becomes closer to the x-axis. Hence $\log(x)$ rises slower than $\ln(x)$. Both have the same x-intercept of $(1, 0)$.

2) Both are reflections over the line $y = x$

A point (x, y) becomes (y, x) on the other curve

The y-axis is an asymptote for the log curve, while the x-axis is an asymptote for the exponential.

3) Domain is $x > 0$. Range is all real numbers.

4) (a) 6^x

(b) e^x

(c) $\log(x)$

(d) $\log_\pi(x)$

5) $f(g(x)) = 5 \cdot 2^{\wedge}(\log_2(x/5)) = 5*(x/5) = x$

$g(f(x)) = \log_2(5 \cdot 2^x/5) = \log_2(2^x) = x$

6) (a) $\log_2 16 = 4$

(b) $\log_{15} 225 = 2$

(c) $\log 1000 = 4$

(d) $\log_8 2 = 1/3$

6) (e) $\log_{64} 8 = 1/2$

(f) $\ln 1 = 0$

(g) $e^5 = x$

(h) $10^2 = 100$

(i) $3^4 = 81$

(j) $10^0 = 1$

(k) $e^0 = 1$

(l) $5^0 = 1$

7) (a) 3

(b) -3

(c) 1

(d) no solution

(e) -2

(f) 27

8) (a) $x = 3^{27}$

(b) $x = \log 2.5$

(c) $x = (e + 4)/2$

9) (a) $\log_7 x^4 + \log_7 y^4 - \log_7 z^8$

(b) $2 + \log(p + 1) - \log(p) - \log(2 - p)$

(c) $4(\ln z)^2$

10) (a) $\log(x^2 z^2/y^5)$

(b) $\ln(5x^3/(6x^3 e)) = \ln(5/6 \cdot e)$

(c) $\log_5((3 - x)^4/x^5)$

(d) $\log((3 - x)^4/x^5) / \log 5$

11) (a) $\dfrac{\log_2(x^4)}{\log_2(7)} + \dfrac{\log_2(y^4)}{\log_2(7)} - \dfrac{\log_2(z^8)}{\log_2(7)}$

(b) $1 + \log_2(p + 1)/\log_2 10 + \ldots$

(c) $4(\log_2 z/\log_2 e)^2$

12) $\log_b x^n = \log_b(\underbrace{x \cdot x \cdot x \cdot x \cdot \ldots \cdot x}_{n\ \text{times}})$

$= \underbrace{\log_b x + \log_b x + \log_b x + \ldots \log_b x}_{n\ \text{times}}$

$= n \cdot \log_b x$

13) log A + log B + log C + log D – log R – log S
- log T – log V

14) log x = lnx/ln10 = 1/ln10 * lnx = ln(x)^(1/ln 10)
Since 1/ln10 = (ln 10)$^{-1}$, we have ln(x)^(ln(10)^-1))

15)   (a) x = 1.56
    (b) x = 3.347
    (c) x = 2.1

16)   (a) x = 36.706
    (b) x = .5
    (c) x = 23
    (d) x = -2.4
    (e) x = ($10^{D/A}$ – C)/B

17) ? = x·ln2

18)   (a) scatter plot
    (b) y = -78.5 + 18.2 ln(x)
    (c) y = -78.5 + 18.2 ln(400) ~ 30 = 2016
    (d) The r-value is fairly high (r = 0.94), however trying to predict 30 years into the future is
very unrealistic.

19) π(100) ~ 22. The actual answer is 25.

20) 11.461

BOOKLET 10 – The Quadratic Family

1) (a) yes
    (b) yes
    (c) yes
    (d) no, linear
    (e) no, x is to the 3rd power
    (f) no, negative exponent

2) (a) (0,0)
    (b) (4,-1)
    (c) (0.5,0.25)
    (d) (-.786, -.414)
    (e) (0,4)
    (f) $(-e/(2\pi), (-e^2 + 4\alpha\pi)/(4\pi))$

3) (a) x = 0
    (b) x = 4
    (c) x = 0.5
    (d) x = -.786
    (e) x = 0
    (f) $x = -e/(2\pi)$

4) $y = 3x^2 - 4x - 1$

5) $y = 4x^2 + 2x - 7$

6) See Group Pause, page 310

7) Opens down

8) $y = ax^2 + bx + 12$, where a > 0

9) $y = ax^2$, where a < 0

10) (a) 3
    (b) 1
    (c) 4
    (d) 2

11) Use TI to verify solutions to question 2

12) (a) (x + 9)(x – 9)
    (b) $(x + 4)^2$
    (c) (x + 5)(x – 3)
    (d) (2x + 10)(2x – 10) or 4(x + 5)(x – 5)
    (e) $2(x + 4)^2$
    (f) (x + 4)(x – 1)
    (g) 4x(x + 2)
    (h) 2(x + 9)(x + 2)
    (i) 5(x + 1)(x – 1)
    (j) (x + √14)(x - √14)
    (k) $(x + 6)^2$
    (l) A(x + √(B/A))(x - √(B/A))

13) (a) x = $\pm$ 9
    (b) x = -4
    (c) x = {-5, 3}
    (d) x = $\pm$ 5
    (e) x = -4
    (f) x = {1, -4}
    (g) x = {0, -2}
    (h) x = {-2, -9}
    (i) x = $\pm$ 1
    (j) x = $\pm$ √14
    (k) x = -6
    (l) x = $\pm$ √(B/A)

14) x = -b/(2a) $\pm$ √(b^2 – 4ac)/(2a)

15) (a) x = 2 $\pm$ √7
    (b) x = -3 $\pm$ 2√5
    (c) x = 5/4 $\pm$ √(129)/4
    (d) x = 2/a $\pm$ √(4 – 10a)/a

16) See question 14

17) (a) x = 2/5 $\pm$ √19/5
    (b) x = {0.138, 3.61}
    (c) x = $\pm$ 3/2
    (d) x = {0.365, -4.11}
    (e) x = -7/4 $\pm$ √73/4

17)   (f) x = {-5, 3}
      (g) x = 7/4 $\pm$ $\sqrt{57}$/4
      (h) x = -β/(2α) + $\sqrt{(\beta^2 - 4\alpha\gamma)}$/(2$\alpha$)

18) Use TI to verify solutions to question 17

19) ϕ = (1 + $\sqrt{5}$)/2 ~ 1.618

20) See 14

21)   (a) x = 2/5 $\pm$ $\sqrt{11}$/5 i
      (b) x = -.125 $\pm$ 0.696i
      (c) x = $\pm$ 3/2i
      (d) x = -1.875 $\pm$ .992i
      (e) x = -7/4 $\pm$ $\sqrt{71}$/4 i
      (f) x = 1 $\pm$ $\sqrt{14}$ i

22)   (a) y = .882x^2 + 20.623x + 9.99, r = .993
      (b) 775.25 feet
      (c) 267.04 feet

23) x = {-9, 7}

24) f(x) = -x^2 + 9x + 4

25) {8.36, -2.45} or {-8.36, 2.45}

26) {55, 57} (or -55, -57)

27) {24, 26}

28) {7, 38}

BOOKLET 11 – Higher Order Polynomial Family

1)　　(a) no, z
　　　　(b) no
　　　　(c) yes, 2
　　　　(d) no
　　　　(e) no
　　　　(f) yes, 6

2) Answers may vary. General answer is $as^4 + bs^3 + cs^2 + ds + e$ $(a \neq 0)$

3)　　(a) $x = \{0,0,5,-7\}$
　　　　(b) $x = \{0,1,9\}$
　　　　(c) $x = \{\pm 2, \pm 2i\}$
　　　　(d) $x = \{\pm\sqrt{5}, \pm\sqrt{3}i\}$
　　　　(e) $x = \{0,6,6\}$
　　　　(f) $x = .676 \pm 1.534i$

4)　　(a) $x^2 + 5 + 5/(x - 3)$
　　　　(b) $x^2 + 2x - 4 + 7/(x + 2)$

5)　　(a) $x = \{1, 5 \pm i\}$
　　　　(b) $x = \{-4, 2 \pm i/2\}$

6)　　(a) $x = \{\pm 1.909, \pm 1.28i\}$
　　　　(b) $x = -.225$
　　　　(c) $x = \{.0039, 10.307\}$
　　　　(d) $x = \{.0498, 20.086\}$ or $\{e^3, 1/e^3\}$

7)　　(a) $x^4 - 3x^3 - 4x^2$
　　　　(b) $x^3 - 5x^2 + 8x - 6$
　　　　(c) $x^3 + 2x^2 + 4x + 8$
　　　　(d) $x^4 - 1$

8) (a) $a^4 - 4a^3b + 6a^2b^2 - 4ab^3 + b^4$
　　(b) $x^5 + 15x^4 + 90x^3 + 270x^2 + 405x + 243$
　　(c) $y^4 - 4y^3x + 6y^2x^2 - 4yx^3 + x^4$
　　(d) $64 p^6 + 576 p^5 q + 2160 p^4 q^2 + 4320 p^3 q^3 + 4860 p^2 q^4 + 2916 p q^5 + 729 q^6$